EVOLUTION

EVOLUTION
The Story of Life on Earth

Written by JAY HOSLER

Art by KEVIN CANNON and ZANDER CANNON

A Novel Graphic from Hill and Wang

A division of Farrar, Straus and Giroux

New York

Hill and Wang
A division of Farrar, Straus and Giroux
18 West 18th Street, New York 10011

This is a Z FILE, INC. Book
Distributed in Canada by D&M Publishers, Inc.
Printed in the United States of America
Published in 2011 by Hill and Wang
First paperback edition, 2012

The Library of Congress has cataloged the hardcover edition as follows:
Hosler, Jay (Jay S.)
    Evolution : the story of life on Earth / written by Jay Hosler. -- 1st ed.
       p. cm.
    ISBN : 978-0-8090-9476-9 (hardcover : alk. paper)
    1. Evolution (Biology) -- Popular works.   I. Title.

QH367.H675 2011
576.8 -- dc22
                                                        2010005777

Paperback ISBN: 978-0-8090-4311-8

Editor: Howard Zimmerman
Design: Kevin Cannon and Zander Cannon

www.fsgbooks.com

5 7 9 10 8 6

For Boo-Boo, Rattley von Stinkenstein, and Hyperbole Queen
--Jay Hosler

For H.L.
--Kevin Cannon

To Julie and Jin-seo
--Zander Cannon

If you ever get a chance to make a book with brilliant, talented, kind people, be sure to take it. I did. Kevin Cannon and Zander Cannon are two of my favorite cartoonists ever. Their talent and robust scientific literacy took the simple genetic code of this script and brought it to life in stunning and unexpected ways. Howard Zimmerman was the best editor and collaborator a writer could ask for. He was usually the smartest guy in the room, and his patience, vision, and knowledge made us all better creators. Finally, I am grateful to Thomas LeBien at Hill and Wang for seeing the tremendous potential of comics and giving me a chance to write a Novel Graphic about the most exciting topic in natural science.

--Jay Hosler

We would like to thank Thomas LeBien and Howard Zimmerman for their support and hard work, as well as Jay Hosler for his impeccable scripting and expertise. Thanks go as well to the Hill and Wang production staff for a streamlined and trouble-free process. Thanks to our families for love and support, and thanks to all who know that seeking out truth, facts, and honesty in no way diminishes the wonder of the world around us.

--Kevin Cannon and Zander Cannon

CONTENTS

N
C
S
O
H
P

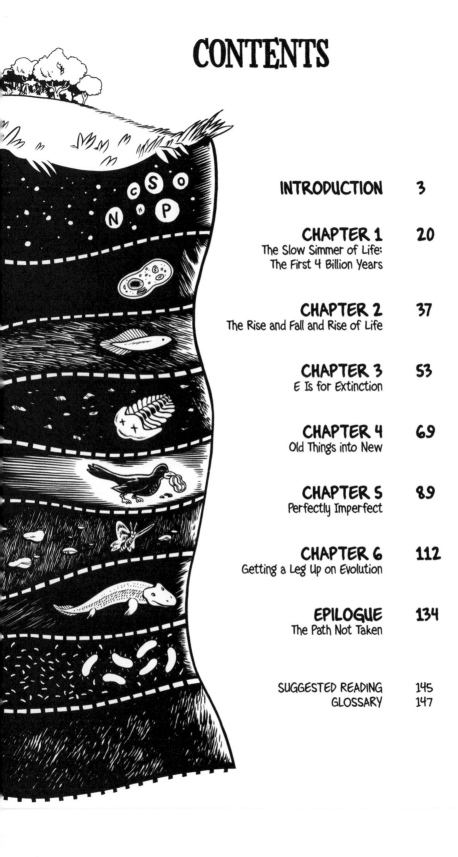

EVOLUTION

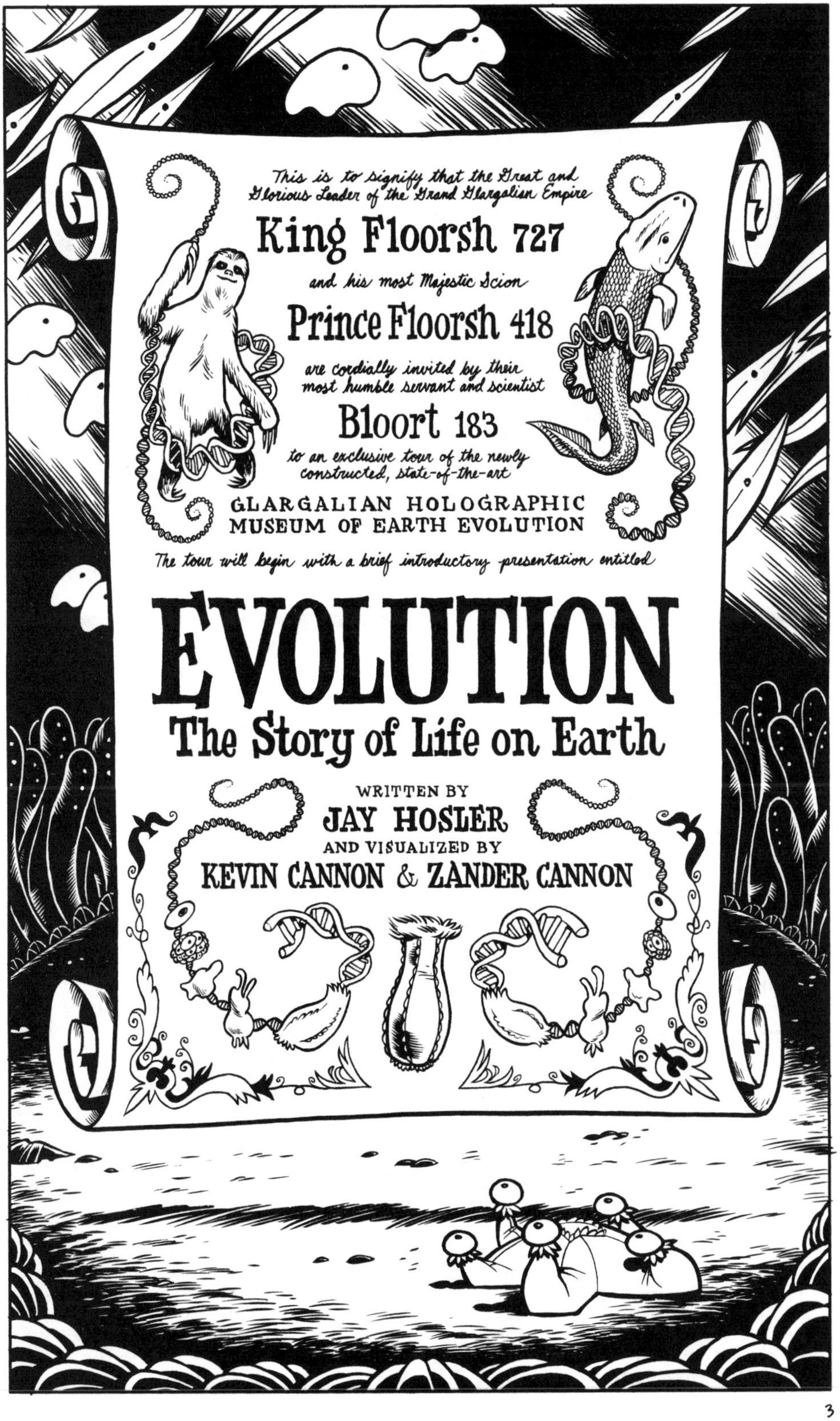
This is to signify that the Great and Glorious Leader of the Grand Glargalian Empire
King Floorsh 727
and his most Majestic Scion
Prince Floorsh 418
are cordially invited by their most humble servant and scientist
Bloort 183
to an exclusive tour of the newly constructed, state-of-the-art
GLARGALIAN HOLOGRAPHIC MUSEUM OF EARTH EVOLUTION
The tour will begin with a brief introductory presentation entitled
EVOLUTION
The Story of Life on Earth
WRITTEN BY
JAY HOSLER
AND VISUALIZED BY
KEVIN CANNON & ZANDER CANNON

EARTH FIRST CAME TO THE ATTENTION OF GLARGALIAN ASTRONOMERS BECAUSE OF THE COLOR IT REFLECTS.

IT'S... GREEN... AND BLUE.

GREAT GLARGALLY-MARGALLY!!

WHILE THIS MAY NOT SOUND VERY EXCITING, TO OUR SCIENTISTS IT SUGGESTED A WORLD THAT MIGHT HAVE BLUE OCEANS AND GREEN PLANTS.

LIFE!

THE FIRST LIFE DISCOVERED BEYOND GLARGAL.

A CLOSER INSPECTION REVEALED AN EXUBERANT VARIETY OF LIVING THINGS. **WHERE** DID THEY ALL COME FROM? **WHAT** WAS THE SOURCE OF THIS INCREDIBLE DIVERSITY?

THE ANSWER, IN PART, IS THAT LIFE ON EARTH IS **TOUGH**. THERE ISN'T ENOUGH WATER, FOOD, AND SPACE FOR EVERY ORGANISM, SO THERE IS A **STRUGGLE TO SURVIVE**.

LOOKS LIKE **ASTEROIDS** TODAY, DEAR.

GOOD THING I BROUGHT AN **UMBRELLA**.

BECAUSE OF THIS, ORGANISMS ON EARTH ARE FORCED TO EVOLVE INNOVATIVE SOLUTIONS TO VARIOUS THREATS TO THEIR EXISTENCE. CHALLENGES TO SURVIVAL CAN COME FROM ANYWHERE, EVEN OUTER SPACE.

DURING ITS LONG HISTORY, EARTH HAS UNDERGONE DRAMATIC CHANGES. THE VERY CONTINENTS HAVE SHIFTED. ISLANDS AND OCEANS HAVE COME AND GONE. THE PLANET HAS PERIODICALLY FROZEN AND THAWED.

THE PRIMORDIAL EARTH WAS SO HOT AFTER IT FORMED THAT ALL THE ROCKS WERE LIQUEFIED. IT REMAINED THIS WAY FOR ABOUT 500 MILLION YEARS, UNTIL THINGS COOLED DOWN AND A HARD OUTER LAYER, CALLED A **CRUST**, FORMED.

AT ABOUT THE SAME TIME, A THIN LAYER OF GASSES WERE TRAPPED BY EARTH'S GRAVITY. THIS FORMED THE PLANET'S ORIGINAL ATMOSPHERE, A NOXIOUS MIX OF METHANE, AMMONIA, AND HYDROGEN.

THE HEAT AND CHEMICALS COMBINED IN A PREHISTORIC STEW. SIMPLE INORGANIC COMPOUNDS SLOSHED AROUND ON THE PLANET'S SURFACE, BUBBLED UP FROM BELOW, AND DROPPED FROM THE SKY ON METEORS.

MOLECULES MIXED AND REACTED, FORMING COMPLEX CHAINS. SOME OF THESE COULD MAKE COPIES OF THEMSELVES. THEY WERE THE FORERUNNERS OF GENETIC MOLECULES LIKE RNA AND DNA...

...THE VERY STUFF OF LIFE!

RIBOZYME

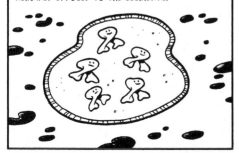

MAKING COPIES OF YOURSELF IN A BIG, BUBBLING STEW CAN BE A CHALLENGE. SOME OF THESE EARLY SELF-COPYING MOLECULES FOUND REFUGE INSIDE OF WATERY BAGS WITH THIN, FATTY MEMBRANES.

THESE EARLY **CELLS** PROVIDED A STABLE ENVIRONMENT FOR THE GENETIC MATERIAL AND GAVE THEM A BIG ADVANTAGE OVER THOSE MOLECULES THAT REMAINED EXPOSED TO THE ELEMENTS.

IN ADDITION TO PASSING THEIR GENETIC MATERIAL TO THEIR OFFSPRING, THESE EARLY CELLS ALSO FREELY SWAPPED GENETIC MATERIAL WITH OTHER, UNRELATED CELLS. THESE EXCHANGES ARE CALLED **LATERAL TRANSFERS** AND THEY PROMOTE VARIETY WITHIN POPULATIONS.

TRADE YA.

DEAL.

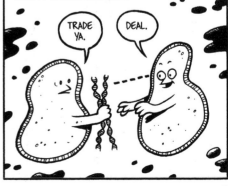

THE THREE DOMAINS OF MODERN EARTH LIFE
EMERGED FROM THIS PRIMITIVE CELLULAR COMMUNITY:
BACTERIA, EUKARYOTES, AND **ARCHAEA.**

BACTERIA

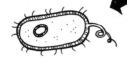

EUKARYOTES

ARCHAEA

BACTERIA ARE ALL SINGLE-CELLED ORGANISMS THAT HAVE A CELL WALL AND A SINGLE, CIRCULAR RING OF DNA. THEY BECAME THE MOST SUCCESSFUL ORGANISMS OF THE PLANET. ONE SPECIES, CALLED CYANOBACTERIA, FILLED EARTH'S EARLY ATMOSPHERE WITH OXYGEN THROUGH **PHOTOSYNTHESIS.**

EUKARYOTES ARE QUITE DIFFERENT FROM BACTERIA. INSTEAD OF HAVING A RIGID CELL WALL, EUKARYOTIC CELLS ARE ENCASED IN A MORE PERMEABLE MEMBRANE. AND EACH HAS AN INTERNAL COMPARTMENT CALLED A **NUCLEUS** TO HOUSE ITS DNA.

THE **ARCHAEA** ARE SINGLE-CELLED ORGANISMS THAT HAVE A MIX OF BACTERIAL AND EUKARYOTIC FEATURES.

SINGLE-CELLED

MULTICELLED

MANY EUKARYOTES REMAINED SINGLE-CELLED ORGANISMS, LIKE PARAMECIA AND AMOEBAS, AND ARE KNOWN AS **PROTISTS.**

BUT ABOUT 1.5 BILLION YEARS AGO, A FEW BRANCHES OF THE EUKARYOTE FAMILY TREE EVOLVED TO LIVE AND WORK TOGETHER AS **MULTICELLED** ORGANISMS. THIS INNOVATION GAVE RISE TO THREE NEW LINES OF LIVING CREATURES: **PLANTS, FUNGI,** AND **ANIMALS.**

EACH WOULD EVOLVE A DIZZYING ARRAY OF SPECIALIZED FEATURES, CALLED **ADAPTATIONS,** TO HELP THEM FIND FOOD, WATER, AND MATES.

PLANTS

ANIMALS

FUNGI

GREEN PLANTS CAN MAKE SUGARS FROM SUNLIGHT, WATER, AND CARBON DIOXIDE, GIVING OFF OXYGEN AS A BY-PRODUCT. THIS PROCESS IS CALLED PHOTOSYNTHESIS. THEY GROW IN A FIXED LOCATION AND DON'T MOVE AROUND MUCH.

FUNGI ARE DECOMPOSERS THAT GROW ON OTHER ORGANISMS -- USUALLY DEAD ONES, BUT NOT ALWAYS... THEY ALSO AREN'T VERY MOBILE.

ANIMALS ARE HIGHLY MOBILE AND EAT OTHER ORGANISMS TO GET THEIR FOOD. THIS LINE WOULD PRODUCE SPECIES CAPABLE OF THOUGHT.

YES, PRINCE FLOORSH 418. ALTHOUGH THE SPECIES WITH THE MOST SOPHISTICATED INTELLECT ON EARTH IS VERY DIFFERENT FROM US, AS WE WILL SOON SEE. ...ASIDE FROM THE OBVIOUS FACT THAT WE ARE **OCEAN-DWELLERS** AND THE HUMANS ARE **LAND-DWELLERS**.

So, Bloort...I assume this report is as long -- that is, as **THOROUGH** -- as your initial one on human genetics?

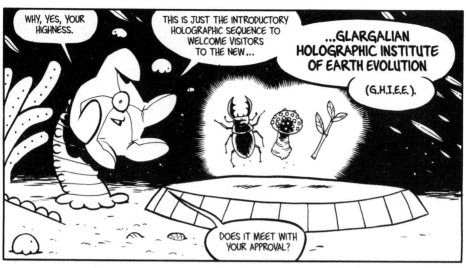

WHY, YES, YOUR HIGHNESS.

THIS IS JUST THE INTRODUCTORY HOLOGRAPHIC SEQUENCE TO WELCOME VISITORS TO THE NEW...

...GLARGALIAN HOLOGRAPHIC INSTITUTE OF EARTH EVOLUTION

(G.H.I.E.E.).

DOES IT MEET WITH YOUR APPROVAL?

It's fine, I guess. It isn't exactly SQUINCHLAND FUN PARK, but --

I think it's GREAT. Just look at all of these weird creatures.

Can we keep going, Dad? Can we?

Well, certainly, if it pleases you, my son.

ANY INPUT YOU AND THE PRINCE MIGHT HAVE WOULD BE DEEPLY APPRECIATED, YOUR MAJESTY. I WOULD LIKE TO WORK OUT ALL OF THE BUGS BEFORE WE OPEN THE MUSEUM TO THE PUBLIC.

Not all the bugs, I hope. These things are COOL.

THE EVOLUTION OF LIFE ON EARTH BEGAN IN ITS SEAS, BUT ABOUT 450 MILLION YEARS AGO A FULL-SCALE INVASION OF THE LAND BEGAN.

WITHIN 100 MILLION YEARS, EARTH WAS COVERED IN SPECIES THAT OOZED, SLITHERED, BLOOMED, BURROWED, AND TOOK TO THE AIR.

QUITE RECENTLY, LESS THAN A QUARTER-MILLION YEARS AGO, A TERRESTRIAL SPECIES EVOLVED THAT WOULD DISTINGUISH ITSELF LIKE NO OTHER BY RESHAPING THE WORLD WITH TWO STUBBY DIGITS CALLED OPPOSABLE THUMBS.

NOW, **THIS** IS WHAT I CALL AN ADAPTATION.

WITH THEIR GRASPING HANDS AND THEIR BIG BRAINS, THE HUMANS CREATED LANGUAGE, ART, AND SCIENCE. THEY DISCOVERED WAYS TO DESCRIBE THE NATURAL WORLD AND THE WORLD OF PURE ABSTRACTION.

Wow! They sound fascinating. Are we going to hear more about these creatures?

What do you mean? Bloort covered much of this in his last report on Earth. Didn't you read it?

Well, I...uh... started it.

I WILL BE HAPPY TO REVIEW ANYTHING THE PRINCE DESIRES, SIRE.

Hmph. The Prince should **DESIRE** to be more thoroughly prepared if he ever hopes to be king.

ABOUT THE **HUMANS**, THEN...

THEY MADE TOOLS THAT INCREASED THEIR CHANCES OF SURVIVAL. AND THEY MANIPULATED EVOLUTIONARY PROCESSES TO TURN WILDLIFE INTO DOMESTICATED LIVESTOCK AND FOOD PLANTS.

THEY POKED, PROBED, AND PONDERED THEIR WORLD. IN TIME, THEY LEARNED TO READ THE VERY ROCKS THEMSELVES. FOSSILS OFFERED THEM A GLIMPSE INTO WORLDS AND CREA- TURES THAT HAD COME AND GONE LONG BEFORE THEIR TIME.

STRANGE LIVING SPECIES LIKE BIRDS THAT COULDN'T FLY AND SNAKES WITH LEGS WERE EVIDENCE THAT **EVOLUTION MIGHT BE AN ONGOING, CONTINUOUS PROCESS.**

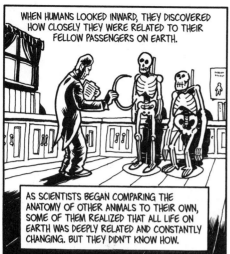

WHEN HUMANS LOOKED INWARD, THEY DISCOVERED HOW CLOSELY THEY WERE RELATED TO THEIR FELLOW PASSENGERS ON EARTH.

AS SCIENTISTS BEGAN COMPARING THE ANATOMY OF OTHER ANIMALS TO THEIR OWN, SOME OF THEM REALIZED THAT ALL LIFE ON EARTH WAS DEEPLY RELATED AND CONSTANTLY CHANGING. BUT THEY DIDN'T KNOW HOW.

Have they figured it out yet?

THEY HAVE. ABOUT 150 YEARS AGO A HUMAN NAMED CHARLES DARWIN PROPOSED A MECHANISM FOR EVOLUTION IN HIS BOOK **ON THE ORIGIN OF SPECIES.**

THE BOOK WAS A SENSATION AND SOLD OUT THE DAY IT WAS PUBLISHED.

PEOPLE WENT "APE" FOR NATURAL SELECTION.

DADUM PTISSHH

Uh...

I don't get it.

OH, SORRY ABOUT THAT. **NATURAL SELECTION** IS THE NAME OF THE EVO-LUTIONARY MECHANISM I PROPOSED.

IT'S THE PROCESS BY WHICH **FAVORABLE TRAITS ARE PRE-SERVED** IN A GROUP OF ORGANISMS AND **HARMFUL TRAITS DIE OUT.** SOME ALSO REFER TO THIS AS "SURVIVAL OF THE FITTEST."

Bloort, why is this strange creature speaking to me?

IT IS A FULLY **INTERACTIVE** HOLOGRAM, YOUR MAJESTY.

So, how did you figure it out, sir?

IT TOOK ME A LONG TIME. IN RETROSPECT, I SUPPOSE I SPENT THE FIRST HALF OF MY LIFE TRAINING FOR THIS DISCOVERY.

AS A BOY, I COLLECTED EVERY-THING FROM ROCKS TO LITTLE BITS OF PLASTER.

O-o-o-o, THIS PEBBLE IS **AWESOME.**

10

MY BROTHER ERASMUS AND I DID CHEMISTRY EXPERIMENTS IN A TOOL SHED ON OUR PROPERTY.

POOF!

A LITTLE LESS **SALTPETER** NEXT TIME.

CHECK.

I DIDN'T CARE FOR SCHOOL MUCH BUT I WAS FASCINATED BY THE NATURAL WORLD. IN FACT, I WAS ONCE SO ENGROSSED IN MY THOUGHTS THAT I WALKED RIGHT OFF A WALL.

MY FATHER LET ME KNOW THAT HE WAS... **CONCERNED**...ABOUT MY LACK OF FOCUS.

YOU CARE FOR NOTHING BUT **SHOOTING**, **DOGS**, AND **RAT-CATCHING**, AND YOU WILL BE A DISGRACE TO YOURSELF AND ALL YOUR FAMILY.

RAT CATCHER'S DIGEST

FRANKLY, I THOUGHT THAT WAS A LITTLE UNFAIR. BUT HE WAS DETERMINED TO WHIP ME INTO SHAPE, SO HE SENT ME TO MEDICAL SCHOOL. WHEN THAT DIDN'T TAKE, HE SENT ME TO CAMBRIDGE UNIVERSITY TO BECOME A MINISTER.

You became a **MINISTER**?

NOT QUITE.

JUST BEFORE I TOOK MY FINAL TESTS, I GOT THE OPPORTUNITY TO SAIL AROUND THE WORLD FOR FIVE YEARS.

AS THE **SHIP'S NATURALIST** ON THE HMS BEAGLE, I COULD FINALLY APPLY MY LOVE OF SHOOTING, DOGS, AND RAT-CATCHING. I SENT CRATE AFTER CRATE OF BIOLOGICAL WONDERS BACK TO ENGLAND.

BY THE TIME I RETURNED, FIVE YEARS LATER, I HAD ALREADY MADE A NAME FOR MYSELF. THEN I GOT MARRIED, HAD KIDS, AND RAN EXPERIMENTS FOR THE NEXT TWENTY YEARS.

THE FRUIT OF MY RESEARCH WAS A MECHANISM FOR EVOLUTION CALLED NATURAL SELECTION. THIS MECHANISM IS SO SIMPLE THAT, WHEN I PUBLISHED MY RESULTS, THE FAMOUS ENGLISH BIOLOGIST **THOMAS HENRY HUXLEY** SAID --

HOW **STUPID** OF ME NOT TO HAVE THOUGHT OF THAT!

So, are you gonna tell us how it works, or not?

MAY I?

IT'S YOUR THEORY.

BRILLIANT. SO, THERE ARE FOUR BASIC CONDITIONS THAT MUST BE MET FOR NATURAL SELECTION TO OCCUR.

1 THE FIRST IS THAT **TRAITS IN A POPULATION OF ORGANISMS EXHIBIT VARIATION.**

IN OTHER WORDS, EVERYBODY IS SLIGHTLY DIFFERENT.

A QUICK LOOK AROUND WILL CONFIRM THAT MEMBERS OF ANY GIVEN SPECIES AREN'T EXACTLY ALIKE. INDIVIDUALS CAN VARY IN TRAITS THAT RANGE FROM SIZE AND SHAPE TO HOW THEY ACT OR HOW THEY FUNCTION CHEMICALLY.

 **THE SECOND IMPORTANT CONDITION IS THAT IN ANY GIVEN POPULATION, NOT ALL INDIVIDUALS SURVIVE TO REPRODUCE.**

THIS IS TRUE FOR EVERY ORGANISM. FROGS, FOR EXAMPLE, MAY PRODUCE THOUSANDS OF EGGS, BUT ONLY A HANDFUL LIVE LONG ENOUGH TO BECOME ADULT FROGS.

AND EVEN IF AN ANIMAL DOES MAKE IT TO ADULTHOOD, THAT IS NO GUARANTEE THAT IT WILL REPRODUCE, ESPECIALLY IF IT CAN'T FIND SOMEONE WILLING TO MATE WITH IT.

THE THIRD CONDITION STATES THAT SURVIVAL IS NOT RANDOM. SURVIVORS MUST HAVE AN ADVANTAGE OVER THOSE THAT DON'T SURVIVE.

THIS ADVANTAGE IS THE RESULT OF A PARTICULAR **VARIATION** IN A **TRAIT,** SUCH AS A SLIGHTLY LONGER TONGUE.

FINALLY, THE FOURTH CONDITION THAT MUST BE MET IS: THE SURVIVOR'S ADVANTAGEOUS TRAITS MUST BE HERITABLE. EVOLUTION CAN HAPPEN ONLY IF ADVANTAGES CAN BE PASSED FROM ONE GENERATION TO THE NEXT.

UNDER THESE CONDITIONS, SOME INDIVIDUALS CONTRIBUTE MORE OFFSPRING TO THE NEXT GENERATION AND THEIR TRAITS WILL SLOWLY GROW MORE AND MORE PREVALENT IN THE POPULATION. **WHEN THIS HAPPENS, THE POPULATION IS EVOLVING.**

IT IS A BEAUTIFUL THEORY, IF I DO SAY SO MYSELF. IT WOULD HAVE ALL BEEN PERFECT IF I COULD HAVE WORKED OUT THE PUZZLE OF INHERITANCE.

INDEED.

THE PREVAILING IDEA AT THE TIME WAS THAT OFFSPRING WERE SIMPLY A BLEND -- AN **AVERAGING** -- OF THEIR PARENTS' FEATURES.

HMM. THAT CAN'T BE RIGHT...

BUT, IF THIS WERE THE CASE, ANY ADVANTAGES THAT MIGHT LEAD TO EVOLUTIONARY CHANGE COULD NEVER ACCUMULATE. THEY'D JUST GET BLENDED AWAY.

DARWIN DIED NEVER KNOWING THAT, AT THE VERY TIME HE WAS STRUGGLING WITH THE CONCEPT OF INHERITANCE, A MONK NAMED GREGOR MENDEL WAS UNLOCKING ITS SECRETS...

...USING PEAS TO DISCOVER THE **UNITS OF INHERITANCE** WE CALL **GENES.**

Ah -- I remember "genes" from your initial report on Earthly genetics. If I recall correctly, they are made of **DNA.**

No way! It all comes back to those self-replicating molecules from the dawn of time!?!

IT IS A DELIGHT TO BASK IN SUCH PERCEPTIVENESS, PRINCE FLOORSH 418.

THEY ARE, IN FACT, AT THE HEART OF ALL OF THIS EVOLUTIONARY CHANGE.

LET'S TAKE A LOOK AT NATURAL SELECTION IN LIGHT OF WHAT WE KNOW ABOUT GENETICS. THE INSTRUCTIONS WERE PRETTY SIMPLE FOR THE FIRST SELF-REPLICATING MOLECULES:

I JUST NEED TO COPY MYSELF.

BUT, AS THEY EVOLVED INTO MORE AND MORE COMPLEX ORGANISMS, THE INSTRUCTIONS ALSO BECAME MORE COMPLEX.

WHOA, THIS LOOKS **COMPLICATED...**

TREE
BLUEPRINTS

Wait, I'm confused. Where are these new DNA instructions coming from?

Mutations and the recombination of genes, if I recall correctly from Bloort's first report.

YOUR REGAL MEMORY IS AS BOUNDLESS AS EVER, SIRE.

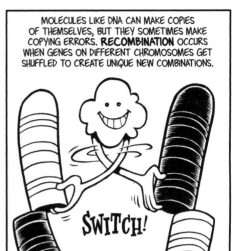

MOLECULES LIKE DNA CAN MAKE COPIES OF THEMSELVES, BUT THEY SOMETIMES MAKE COPYING ERRORS. **RECOMBINATION** OCCURS WHEN GENES ON DIFFERENT CHROMOSOMES GET SHUFFLED TO CREATE UNIQUE NEW COMBINATIONS.

SWITCH!

A **MUTATION** OCCURS WHEN A GENE'S CODE IS COPIED INCORRECTLY. SOME OF THE TIME, THIS CHANGES THE CODE IN A WAY THAT PREVENTS AN ORGANISM FROM GROWING OR FUNCTIONING PROPERLY. THESE ARE CALLED **LETHAL** MUTATIONS.

NO FOOLIN'...

BUT SOMETIMES A MUTATION CAUSES NO HARM, IN WHICH CASE IT IS A NEUTRAL MUTATION. OR IT MAY ACTUALLY **IMPROVE** THE WAY THE ORGANISM WORKS AND IS CALLED A BENEFICIAL MUTATION.

YOU GUYS SEE THAT?

SEE WHAT?

WHAT'S "SEE"?

How do they know when to mutate?

THEY DON'T, YOUR HIGHNESS. MUTATIONS OCCUR AT RANDOM. THEY'RE NOT PLANNED.

You're telling me all of this exquisite evolutionary change is entirely the product of random events?

NOT ENTIRELY, YOUR HIGHNESS.

But...I ...you just...I'm confused.

15

PERHAPS A LITTLE VISUALIZATION WOULD HELP, SIRE.

Perhaps it would.

FIRST, SOME DEFINITIONS.

HERE WE HAVE THE CHROMOSOMES OF TWO DIFFERENT SPECIES OF BACTERIA.

A CHROMOSOME IS A THREAD-LIKE STRUCTURE OF DNA THAT CONTAINS MULTIPLE GENES.

A GENE IS AN INDIVIDUAL UNIT OF INHERITANCE. EACH GENE CARRIES INSTRUCTIONS FOR BUILDING PARTS OF THE BODY.

THE GENES ARE REPRESENTED BY DIFFERENT TONES.

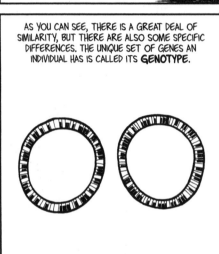

AS YOU CAN SEE, THERE IS A GREAT DEAL OF SIMILARITY, BUT THERE ARE ALSO SOME SPECIFIC DIFFERENCES. THE UNIQUE SET OF GENES AN INDIVIDUAL HAS IS CALLED ITS **GENOTYPE.**

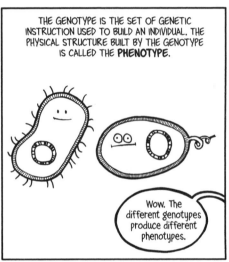

THE GENOTYPE IS THE SET OF GENETIC INSTRUCTION USED TO BUILD AN INDIVIDUAL. THE PHYSICAL STRUCTURE BUILT BY THE GENOTYPE IS CALLED THE **PHENOTYPE.**

Wow. The different genotypes produce different phenotypes.

DIFFERENT INSTRUCTIONS, DIFFERENT BODY.

THE VARIATION THAT DARWIN OBSERVED IN A POPULATION OF PLANTS OR ANIMALS WAS VARIATION IN THEIR **PHENOTYPES** -- THEIR PHYSICAL STRUCTURES.

RANDOM MUTATIONS INTRODUCE NEW GENES THAT PRODUCE SLIGHTLY DIFFERENT PHENOTYPES. SO, MUTATIONS ARE RANDOM...

...BUT WHICH PHENOTYPES SURVIVE IS **NOT.**

THIS CAN BE SEEN QUITE CLEARLY IN A GROWING MEDICAL PROBLEM ON EARTH.

THE FUNGUS *PENICILLIUM NOTATUM* LIVES IN THE SOIL AND COMPETES WITH SINGLE-CELLED BACTERIA FOR RESOURCES. OVER MILLIONS OF YEARS OF COMPETITION, THE FUNGUS HAS EVOLVED THE ADAPTIVE ABILITY TO SECRETE A CHEMICAL THAT KILLS THE COMPETING BACTERIA.

PENICILLIUM NOTATUM

IN 1928, THE HUMAN DOCTOR ALEXANDER FLEMING DISCOVERED THIS CHEMICAL WHEN *PENICILLIUM NOTATUM* KILLED SOME BACTERIAL COLONIES HE WAS TRYING TO GROW.

I SHALL CALL IT... PENICILLIN.

NOW, HUMANS DON'T HAVE THE SAME PROBLEM WITH BACTERIA THAT FUNGI DO. MOST BACTERIA ARE HARMLESS, AND SOME ARE ESSENTIAL FOR HUMAN SURVIVAL. IN THE HUMAN GUT ALONE, THERE ARE HUNDREDS OF BACTERIAL SPECIES THAT AID IN THE DIGESTION OF FOOD.

BEAN BURRITOS AGAIN TONIGHT.

BETTER GET TO WORK AND CRANK OUT SOME GAS.

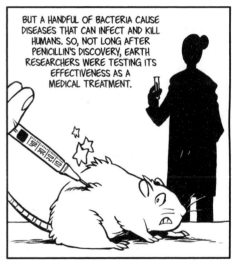

BUT A HANDFUL OF BACTERIA CAUSE DISEASES THAT CAN INFECT AND KILL HUMANS. SO, NOT LONG AFTER PENICILLIN'S DISCOVERY, EARTH RESEARCHERS WERE TESTING ITS EFFECTIVENESS AS A MEDICAL TREATMENT.

LONG STORY SHORT, IT WAS A **WHOPPING SUCCESS!** REMEMBER THE BACTERIAL CELL WALL WE MENTIONED?

Gosh, who could forget?

Uh...

WELL, PENICILLIN WEAKENS THE CELL WALL AND EVENTUALLY CAUSES BACTERIA TO RUPTURE AND DIE.

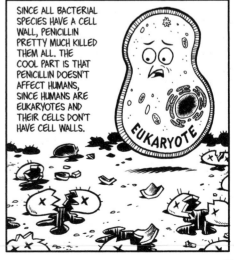

SINCE ALL BACTERIAL SPECIES HAVE A CELL WALL, PENICILLIN PRETTY MUCH KILLED THEM ALL. THE COOL PART IS THAT PENICILLIN DOESN'T AFFECT HUMANS, SINCE HUMANS ARE EUKARYOTES AND THEIR CELLS DON'T HAVE CELL WALLS.

EUKARYOTE

PENICILLIN WAS THE FIRST **ANTIBIOTIC** -- A DRUG THAT TARGETS AND KILLS BACTIERA -- AND IT WAS HAILED AS A "MAGIC BULLET."

BUT OVER THE YEARS, THE "MAGIC" STARTED TO WEAR OFF. IT TOOK MORE AND MORE PENICILLIN TO KILL THE BACTERIA.

THEY WERE EVOLVING.

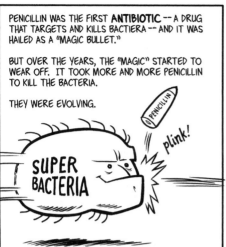

LET'S IMAGINE WHAT WAS HAPPENING IN TERMS OF NATURAL SELECTION. BACTERIAL POPULATIONS DISPLAY VARIATION IN TRAITS JUST LIKE ANY OTHER ORGANISMS. EACH HAS A SLIGHTLY DIFFERENT GENOTYPE AND PHENOTYPE.

WHEN THOSE POPULATIONS WERE ATTACKED WITH PENICILLIN, THE ENVIRONMENT CHANGED AND THE VAST MAJORITY DIED.

BUT A FEW SUR-VIVED. THEY HAD AN ADVANTAGE THE DEAD DID NOT.

UH... WHAT JUST HAPPENED?

A RANDOM MUTATION HAD CHANGED THEIR GENOTYPE, GIVING THEM A NEW PHENOTYPE. THESE MUTANT BACTERIA HAD AN ENZYME THAT CHOPPED UP PENICILLIN BEFORE IT COULD DAMAGE THE CELL WALL.

PRIOR TO THE PRESENCE OF PENICILLIN, THIS MUTATION WAS PROBABLY NEUTRAL. IT DIDN'T HURT, BUT IT DIDN'T HELP. BUT, IN THE FACE OF A PENICILLIN ASSAULT...

...THIS NEW ENZYME WAS DEFINITELY A BENEFICIAL ADAPTATION.

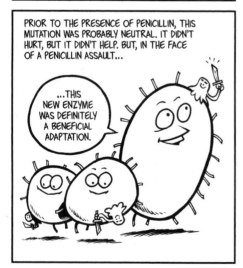

BECAUSE THE GENE FOR **PENICILLIN RESISTANCE** WAS HERITABLE, THERE WERE MORE AND MORE RESISTANT BAC-TERIA IN EACH SUBSEQUENT GENERATION. THE BACTERIA POPULATION WAS EVOLVING!

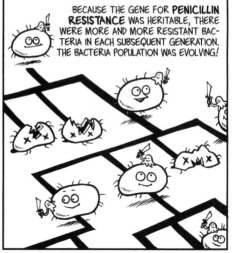

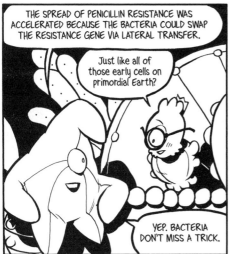

THE SPREAD OF PENICILLIN RESISTANCE WAS ACCELERATED BECAUSE THE BACTERIA COULD SWAP THE RESISTANCE GENE VIA LATERAL TRANSFER.

Just like all of those early cells on primordial Earth?

YEP. BACTERIA DON'T MISS A TRICK.

SO, **RANDOM** MUTATIONS GENERATE THE RAW MATERIAL THAT NATURAL SELECTION THEN WORKS UPON IN A VERY **DIRECTED** FASHION.

And when natural selection favors a particular phenotype for survival, it is also favoring the survival of a particular genotype.

FOR THIS REASON, THE MODERN DEFINITION OF EVOLUTION IS: **A CHANGE IN THE GENETIC MAKEUP OF A POPULATION OVER TIME.**

SUPER BACTERIA

Coool. Hey, Dad, can I go play with the penicillin-resistant bacteria?

It is your royal prerogative.

Sweet!

ER, SIRE --

I'M NOT SO SURE THAT'S SUCH A GOOD IDEA...

Oh, I'm sure he's fine. Now, was this antibiotic resistance the **REAL** reason you wanted to give me this tour?

UH...YES, YOUR MOST PERCEPTIVE HIGHNESS. I WAS HOPING THIS MIGHT PROVIDE US WITH SOME INSIGHT INTO THE HERITABLE DISEASE THAT IS ATTACKING OUR SPECIES, AND...

hee hee --

SQUISH SQUISH

GASP!
The Prince!

EEEEE

ZAP

WE HAVE TO HURRY, SIRE! IT'S SENT HIM TO THE NEXT PART OF THE PRESENTATION: **THE PRECAMBRIAN PERIOD!**

CHAPTER 1

The Slow Simmer of Life: The First 4 Billion Years

By all that is squincheous! He's caught in a **METEOR SHOWER!**

IT'S ALL RIGHT, YOUR HIGHNESS. THEY'RE JUST **HOLOGRAMS.**

THIS IS THE FIRST 500 MILLION YEARS OF EARTH'S EXISTENCE, BACK WHEN IT WAS PUMMELED BY TONS OF **SPACE DEBRIS.**

Well, then, perhaps we should skip ahead a bit. And you, young squinch, will stick with me from now on.

Sorry, Dad. I was just excited to see where it all began.

It's hard to believe that this place was suitable for life, Bloort.

IT **IS** PRETTY NASTY. BUT, BELIEVE IT OR NOT, CONDITIONS WERE JUST RIGHT FOR SOME VERY CREATIVE CHEMISTRY.

SCIENTISTS SUSPECT THAT THE CHEMICAL PRECURSORS FOR LIFE STARTED TAKING SHAPE ABOUT 500 MILLION YEARS AFTER THE EARTH FORMED.

IS THE BOMBARDMENT OVER?

IT LOOKS CLEAR...

AMMONIA

METHANE

THERE'S NOT MUCH OF A FOSSIL RECORD FROM THAT TIME, SO SCIENTISTS HAVE HAD TO TAKE WHAT THEY KNOW ABOUT THE CURRENT CHEMISTRY OF LIFE AND WORK **BACKWARDS.**

TODAY, ALL LIFE ON EARTH WORKS IN THE SAME BASIC WAY. THE DNA FOUND IN EVERY ORGANISM'S CELLS CONTAINS THE INSTRUCTIONS FOR MAKING PROTEINS. THESE PROTEINS THEN --

Wait, wait, wait. What are **PROTEINS?**

They're in his first report, that's what they are.

PROTEINS ARE MOLECULES THAT PERFORM THE FUNCTIONS ESSENTIAL FOR LIFE, PRINCE FLOORSH. SOME CARRY MESSAGES WITHIN A CELL OR BETWEEN TWO DIFFERENT CELLS.

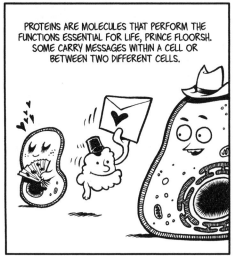

OTHERS ACT AS ANTIBODIES TO DETECT FOREIGN MATERIALS IN AN ORGANISM.

IT'S NO USE HIDING. WE KNOW YOU'RE THERE.

PROTEINS ARE USED TO PROVIDE STRUCTURAL SUPPORT FOR A CELL AS WELL AS TRANSPORTING AND STORING MATERIALS.

ENZYME

ENZYME

BUT PERHAPS THE MOST IMPORTANT FUNCTION OF PROTEINS IS THEIR ABILITY TO ACT AS **ENZYMES**. ENZYMES ARE PROTEINS THAT RUN THE THOUSANDS OF CHEMICAL REACTIONS THAT TAKE PLACE IN A CELL.

How does a mindless protein know which job it's supposed to do?

GOOD QUESTION, YOUR PRINCELINESS.

IF WE UNWIND A PROTEIN WE CAN SEE THAT IT IS A STRING OF SMALLER MOLECULES, CALLED **AMINO ACIDS**, CONNECTED TO EACH OTHER IN A UNIQUE SEQUENCE.

ZRRRRT

21

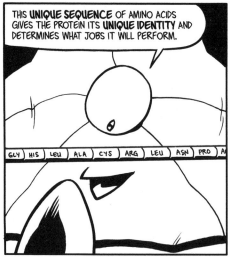

THIS **UNIQUE SEQUENCE** OF AMINO ACIDS GIVES THE PROTEIN ITS **UNIQUE IDENTITY** AND DETERMINES WHAT JOBS IT WILL PERFORM.

GLY | HIS | LEU | ALA | CYS | ARG | LEU | ASN | PRO | A

So, in other words, when you string together specific **AMINO ACIDS**...

...in a specific **ORDER**...

...you're coding for a specific **FUNCTION**.

PRECISELY, YOUR HIGHNESS.

But how does the cell know which amino acids to string together to make a certain protein?

AH! **THAT'S** THE INFORMATION ENCODED IN THE **DNA.**

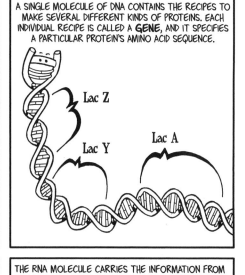

A SINGLE MOLECULE OF DNA CONTAINS THE RECIPES TO MAKE SEVERAL DIFFERENT KINDS OF PROTEINS. EACH INDIVIDUAL RECIPE IS CALLED A **GENE**, AND IT SPECIFIES A PARTICULAR PROTEIN'S AMINO ACID SEQUENCE.

Lac Z

Lac Y

Lac A

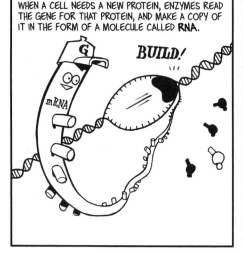

WHEN A CELL NEEDS A NEW PROTEIN, ENZYMES READ THE GENE FOR THAT PROTEIN, AND MAKE A COPY OF IT IN THE FORM OF A MOLECULE CALLED **RNA.**

G

BUILD!

mRNA

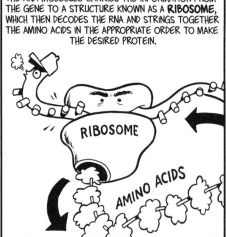

THE RNA MOLECULE CARRIES THE INFORMATION FROM THE GENE TO A STRUCTURE KNOWN AS A **RIBOSOME**, WHICH THEN DECODES THE RNA AND STRINGS TOGETHER THE AMINO ACIDS IN THE APPROPRIATE ORDER TO MAKE THE DESIRED PROTEIN.

RIBOSOME

AMINO ACIDS

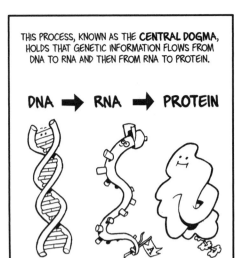

THIS PROCESS, KNOWN AS THE **CENTRAL DOGMA**, HOLDS THAT GENETIC INFORMATION FLOWS FROM DNA TO RNA AND THEN FROM RNA TO PROTEIN.

DNA ➡ RNA ➡ PROTEIN

THE QUESTION IS, HOW DO YOU GET TO THIS NICE, NEAT SYSTEM...

...FROM THIS PRIMORDIAL MESS?

IN 1953, TWO HUMAN SCIENTISTS, STANLEY MILLER AND HAROLD UREY, RAN AN EXPERIMENT TO FIND OUT HOW.

COME IN, COME IN!

MILLER UREY

SIMULATED LIGHTNING

VACUUM

SIMULATED ATMOSPHERE

SIMULATED OCEAN

CONDENSER

HEAT

ELECTRICITY

ICITY

WE WANTED TO SEE IF THE CHEMICAL BUILDING BLOCKS OF LIFE COULD ARISE NATURALLY IN A DEVICE THAT SIMULATED THE ENVIRONMENTAL CONDITIONS OF THE EARLY EARTH.

BACK THEN, THE PLANET'S ATMOSPHERE WAS COMPOSED MAINLY OF METHANE, AMMONIA, AND HYDROGEN, SO THESE WERE THE GASES THAT WE USED IN OUR APPARATUS.

S. MILLER

H. UREY

WE HEATED WATER TO SIMULATE THE HOT ANCIENT OCEAN AND USED ELECTRIC SPARKS AS A STAND-IN FOR LIGHTNING. BOTH CONDITIONS -- A WARM OCEAN AND ELECTRICITY FROM LIGHTNING -- WERE PRESENT ON EARLY EARTH.

ONCE WE HAD THE WHOLE THING SET UP, WE LET IT BUBBLE, CONDENSE, AND REACT FOR A WHOLE WEEK BEFORE WE CHECKED IT.

IN FACT, WE WERE JUST ABOUT TO CHECK IT, SO IF YOU'LL EXCUSE ME FOR A MOMENT.

BY ALL MEANS.

WHEN MILLER AND UREY UNCORKED THE APPARATUS, THEY MADE AN **AMAZING DISCOVERY.**

Holy Moley! Did they create... **LIFE?**

UH...NO, NOT QUITE, YOUR HIGHNESS. BUT THEY **DID** MAKE NUCLEOTIDES AND AMINO ACIDS.

Oh. That's cool, I guess.

What are nucleotides?

DIDN'T I EXPLAIN THAT BEFORE?

Not on this tour, but you did cover it -- **AT GREAT LENGTH** -- in your previous report on human genetics.

I **REALLY** am sorry I didn't read it, Dad.

24

Yes, well, if I recall correctly, nucleotides are the building blocks of which DNA and RNA are made, while amino acids are used to make proteins.

And Miller and Urey created both in their lab?

CORRECT. THE NEXT BIG QUESTION WAS, WHICH ELEMENT OF THE CENTRAL DOGMA-- DNA, RNA, OR PROTEINS -- CAME **FIRST** IN **THE EVOLUTION OF LIFE**?

It had to be DNA, right? You can't build proteins without the DNA instructions.

TRUE. BUT HOW COULD YOU READ THE DNA WITHOUT HAVING THE PROTEIN ENZYMES **FIRST**?

Huh... I don't know.

NOW EXITING MILLER & UREY'S HOLO-LAB

SCIENTISTS SEARCHING FOR THE ANSWER TO THIS PUZZLE LOOKED FOR A MOLECULE THAT COULD STORE INFORMATION LIKE DNA **AND** RUN REACTIONS LIKE PROTEIN ENZYMES.

Well, we know that RNA stores the information it copies from DNA. Can it run reactions, too?

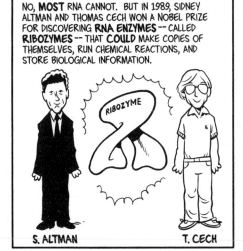

NO, **MOST** RNA CANNOT. BUT IN 1989, SIDNEY ALTMAN AND THOMAS CECH WON A NOBEL PRIZE FOR DISCOVERING **RNA ENZYMES** -- CALLED **RIBOZYMES** -- THAT **COULD** MAKE COPIES OF THEMSELVES, RUN CHEMICAL REACTIONS, AND STORE BIOLOGICAL INFORMATION.

RIBOZYME

S. ALTMAN

T. CECH

RECALL THAT DNA AND RNA ARE SUBJECT TO MUTATION. SINCE **RIBOZYMES** ARE MADE OF RNA, THEY CAN MUTATE AND CHANGE AS WELL.

OOPS, THAT'S NOT RIGHT.

ORIGINAL

COPY

EARTH SCIENTISTS THEORIZE THAT RIBOZYMES COULD EVOLVE IF NATURAL SELECTION FAVORED MUTATIONS THAT ALLOWED SOME TO MAKE FASTER OR MORE STABLE COPIES OF THEMSELVES.

SUCH A MUTATION WOULD ALLOW RIBOZYME MOLECULES TO PRODUCE MORE COPIES OF THEMSELVES THAN THOSE WITHOUT THE MUTATION.

THROUGH A SEQUENCE OF EVENTS STILL NOT CLEAR, SOME RIBOZYMES EVENTUALLY STARTED DOING THEIR COPYING INSIDE OF PRIMITIVE CELL-LIKE STRUCTURES.

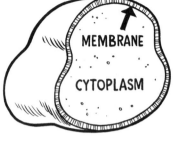

THESE EARLY CELLS CONSISTED OF A THIN **MEMBRANE** FILLED WITH A WATERY SUBSTANCE CALLED **CYTOPLASM**. THEY WEREN'T FANCY, BUT THEY TURNED OUT TO BE A BIG EVOLUTIONARY INNOVATION.

CELLS COMPARTMENTALIZED CHEMICAL REACTIONS AND PROVIDED A STABLE ENVIRONMENT IN WHICH THE RNA COULD FUNCTION.

CELLS ARE ALSO TERRIFIC FOR PUMPING **IN** STUFF THAT THE RNA NEEDS AND PUMPING **OUT** WASTE PRODUCTS.

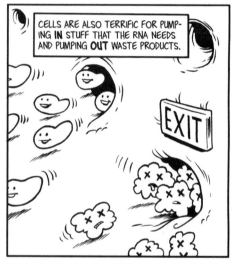

EVENTUALLY, CELLS STARTED STORING GENETIC INFORMATION IN **DNA MOLECULES**, WHOSE DOUBLE-STRANDED STRUCTURE MAKES THEM MORE STABLE THAN RNA.

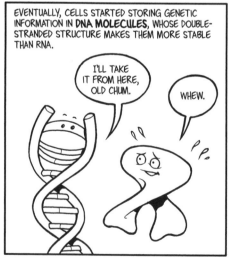

IN THE EARLY STAGES OF THEIR EVOLUTION, CELLS WERE PRETTY SIMILAR AND THEY SWAPPED DNA, RNA, AND PROTEINS FREELY. THESE **LATERAL TRANSFERS** OF GENETIC MATERIAL ADDED VARIATION AND FACILITATED THE SPREAD OF ADAPTIVE TRAITS.

FROM THIS PRIMORDIAL COMMUNITY OF CELLS, THE THREE DOMAINS OF EARTH LIFE EMERGED.

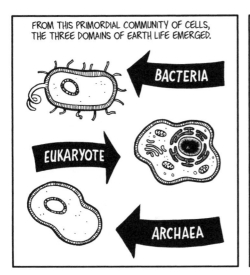

BACTERIA

EUKARYOTE

ARCHAEA

THE BACTERIA AND ARCHAEA ARE COLLECTIVELY CLASSIFIED AS **PROKARYOTES**. PROKARYOTES ARE SINGLED-CELLED ORGANISMS THAT HAVE A CELL WALL SURROUNDING THEIR CELL MEMBRANE...

...THEIR DNA IS A CIRCULAR STRAND THAT FLOATS IN THE CYTOPLASM...

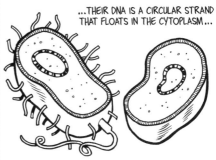

...AND THEY CONTAIN NONE OF THE INTERNAL COMPARTMENTS CALLED ORGANELLES -- WE'LL SEE WHAT ORGANELLES ARE ALL ABOUT SOON ENOUGH.

OF THE THREE DOMAINS, THE FIRST TO HIT IT BIG EVOLUTIONARILY WERE THE **BACTERIA**, WHICH APPEARED ABOUT 3.7 BILLION YEARS AGO.

FIRST CAME THE BEST -- AND THEN CAME ALL THE REST.

BACTERIA WERE THE ONLY FORM OF LIFE ON EARTH FOR ABOUT 2 BILLION YEARS. IN THAT TIME THEY DIVERSIFIED INTO A STUNNING ARRAY OF DIFFERENT SHAPES, SIZES, AND FUNCTIONS.

WHEN LOOKING AT EVOLUTION ON EARTH, THERE ARE TWO SPECIES THAT BEAR SPECIAL CONSIDERATION: **SULFUR-EATING BACTERIA** AND **CYANOBACTERIA**.

THOUGH SULFUR-EATERS AND CYANOBACTERIA EXISTED AT THE SAME TIME, THE **SULFUR-EATERS** WERE THE FIRST **DOMINANT** SPECIES OF BACTERIA. THEY THRIVED IN THE NOXIOUS ATMOSPHERE OF THE EARLY EARTH, CONSUMING **HYDROGEN SULFIDE** -- THE CHEMICAL WE ASSOCIATE WITH THE SMELL OF **ROTTEN EGGS**.

TASTES GREAT...

...BUT IT SURE DOES MAKE MY BREATH STINK.

BUT, THANKS TO CYANOBACTERIA, THE GLORY DAYS OF THE SULFUR-EATERS WERE NUMBERED.

INSTEAD OF EATING HYDROGEN SULFIDE, **CYANO-BACTERIA** HAD EVOLVED THE ABILITY TO COMBINE SUNLIGHT, WATER, AND CARBON DIOXIDE TO CREATE **SUGARS**, FROM WHICH THEY COULD BUILD NEW CELLS. THIS PROCESS -- CALLED **PHOTOSYNTHESIS** -- PRODUCES A HIGHLY REACTIVE WASTE PRODUCT CALLED...

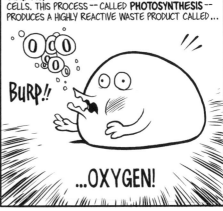

BURP!!

...OXYGEN!

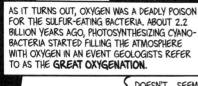

AS IT TURNS OUT, OXYGEN WAS A DEADLY POISON FOR THE SULFUR-EATING BACTERIA. ABOUT 2.2 BILLION YEARS AGO, PHOTOSYNTHESIZING CYANOBACTERIA STARTED FILLING THE ATMOSPHERE WITH OXYGEN IN AN EVENT GEOLOGISTS REFER TO AS THE **GREAT OXYGENATION.**

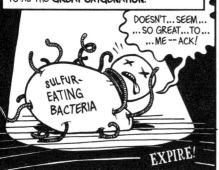

DOESN'T...SEEM... ...SO GREAT...TO... ...ME--ACK!

SULFUR-EATING BACTERIA

EXPIRE!

THE SULFUR-EATING BACTERIA COULD NO LONGER THRIVE AS THEY ONCE HAD AND WERE FORCED TO RETREAT TO EXTREME ENVIRONMENTS, LIKE BOILING-HOT DEEP-SEA VENTS THAT ARE RICH IN SULFUR RATHER THAN OXYGEN.

THEY CAN STILL BE FOUND THERE TODAY.

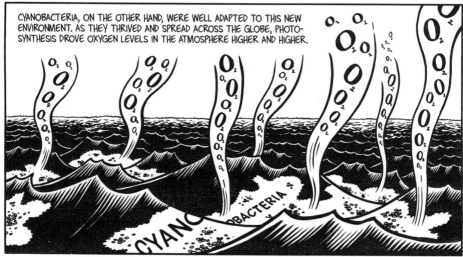

CYANOBACTERIA, ON THE OTHER HAND, WERE WELL ADAPTED TO THIS NEW ENVIRONMENT. AS THEY THRIVED AND SPREAD ACROSS THE GLOBE, PHOTOSYNTHESIS DROVE OXYGEN LEVELS IN THE ATMOSPHERE HIGHER AND HIGHER.

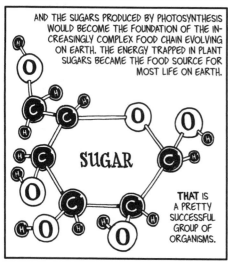

AND THE SUGARS PRODUCED BY PHOTOSYNTHESIS WOULD BECOME THE FOUNDATION OF THE INCREASINGLY COMPLEX FOOD CHAIN EVOLVING ON EARTH. THE ENERGY TRAPPED IN PLANT SUGARS BECAME THE FOOD SOURCE FOR MOST LIFE ON EARTH.

SUGAR

THAT IS A PRETTY SUCCESSFUL GROUP OF ORGANISMS.

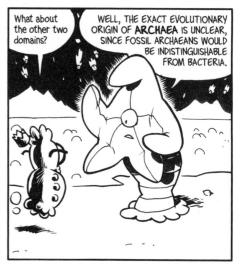

What about the other two domains?

WELL, THE EXACT EVOLUTIONARY ORIGIN OF **ARCHAEA** IS UNCLEAR, SINCE FOSSIL ARCHAEANS WOULD BE INDISTINGUISHABLE FROM BACTERIA.

THE TRUTH IS, THEY WERE DISCOVERED RELATIVELY RECENTLY, AND EARTH SCIENTISTS ARE JUST GETTING TO KNOW THEM.

ONE OF THE MOST INTERESTING THINGS ABOUT ARCHAEANS IS THAT THEY, TOO, ARE CAPABLE OF LIVING IN EXTREME ENVIRONMENTS, SUCH AS NEAR BOILING GEYSERS AND EXTREMELY SALTY WATER.

THE THIRD DOMAIN OF LIFE -- THE **EUKARYOTES** -- FIRST APPEARED ON EARTH ABOUT 1.5 BILLION YEARS AGO AND WOULD BECOME THE SOURCE OF SOME OF EVOLUTION'S MOST INVENTIVE EXPERIMENTS.

WHOA, WHO'S THE NEW GUY?

EUKARYOTIC CELL

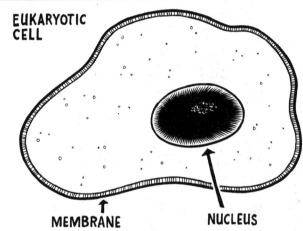

MEMBRANE

NUCLEUS

THE CELLS OF EUKARYOTES ARE QUITE DIFFERENT FROM THE CELLS OF PROKARYOTES. EUKARYOTES HAVE NO CELL WALLS, SO THEIR OUTER MEMBRANE IS THINNER AND MORE FLEXIBLE THAN THE THICK, RIGID CELL WALL OF THE PROKARYOTES.

THEY ALSO HAVE SEVERAL INTERNAL MEMBRANE-BOUND CHAMBERS, CALLED **ORGANELLES**. ONE OF THESE CHAMBERS, CALLED THE **NUCLEUS**, STORES THEIR DNA.

WITH NO CELL WALLS, PRIMITIVE EUKARYOTES USED THEIR PLIABLE CELL MEMBRANES TO EAT OTHER ORGANISMS BY **ENGULFING** THEM.

HEY, DID IT GET **DARK** ALL OF A SUDDEN?

THIS WAS A REMARKABLE ADAPTATION. EATING OTHER ORGANISMS PROVIDED A MORE CONCENTRATED SOURCE OF NUTRIENTS COMPARED TO THE BITS OF ORGANIC MATERIAL PROKARYOTES SUCKED UP FROM THE WATER.

AH, I'M FULL.

MUST BE NICE.

LESS TALKING, MORE ABSORBING, DEAR.

BUT INGESTING OTHERS WOULD ALSO PROVIDE PRIMITIVE EUKARYOTES WITH AN ASTONISHING EVOLUTIONARY OPPORTUNITY.

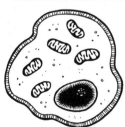

ABOUT 1.5 BILLION YEARS AGO, PRIMITIVE EUKARYOTES ENGULFED ORGANISMS CALLED **ALPHA-PROTEOBACTERIA** THAT WERE VERY GOOD AT UNLOCKING THE ENERGY STORED IN SUGARS.

29

BUT, INSTEAD OF EATING THEM, THE EUKARYOTES EVOLVED A MUTUALLY BENEFICIAL RELATIONSHIP WITH THE ALPHA-PROTEOBACTERIA. THIS INCREASED BOTH OF THEIR CHANCES FOR SURVIVAL.

I'LL PROVIDE THE SHELTER, IF YOU'LL RUN MY METABOLISM.

IT **IS** SPACIOUS...

AND **LOOK** AT THE **VIEW!**

WHEN TWO ORGANISMS BENEFIT BY LIVING TOGETHER IT IS CALLED A **SYMBIOTIC RELATIONSHIP.** IN THIS CASE IT IS AN **ENDOSYMBIOSIS** -- "ENDO" MEANING "INSIDE" -- BECAUSE ONE ORGANISM IS LIVING WITHIN ANOTHER.

THESE ALPHA-PROTEOBACTERIA EVENTUALLY GAVE UP THE ABILITY TO LIVE ON THEIR OWN...

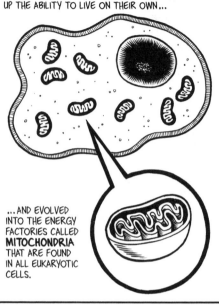

...AND EVOLVED INTO THE ENERGY FACTORIES CALLED **MITOCHONDRIA** THAT ARE FOUND IN ALL EUKARYOTIC CELLS.

ENDOSYMBIOSIS MADE EUKARYOTIC CELLS REMARKABLY EFFICIENT BECAUSE THE HOST CELLS COULD TAKE ADVANTAGE OF SOME COOL BACTERIAL ADAPTATIONS.

AND WHAT CAN **YOU** DO?

I CAN CHANGE SUNLIGHT INTO SUGARY GOODNESS.

YOU'RE IN.

A SECOND ENDOSYMBIOSIS OCCURRED IN THE LINE OF EUKARYOTES THAT WOULD EVOLVE INTO **PLANTS**. THESE CELLS ENGULFED CYANOBACTERIA AND TOOK ADVANTAGE OF THEIR ABILITY TO PHOTOSYNTHESIZE.

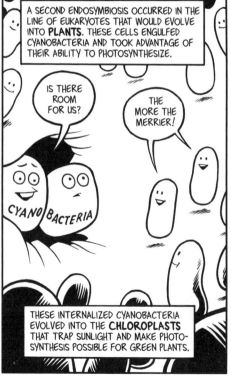

IS THERE ROOM FOR US?

THE MORE THE MERRIER!

CYANOBACTERIA

THESE INTERNALIZED CYANOBACTERIA EVOLVED INTO THE **CHLOROPLASTS** THAT TRAP SUNLIGHT AND MAKE PHOTOSYNTHESIS POSSIBLE FOR GREEN PLANTS.

How did the humans figure that out? I mean, they couldn't see that happen, so what's the evidence that eukaryote cells incorporated bacteria in symbiotic relationships?

PERHAPS YOU CAN TELL ME, YOUR HIGHNESS. WHAT EVIDENCE DO YOU THINK THEY COULD HAVE FOUND?

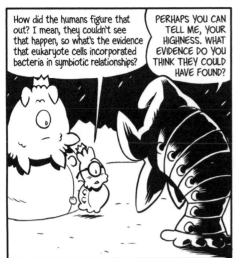

Hmm...Well, you showed us earlier that bacteria all had their own genes.

THAT'S RIGHT. AND MITOCHONDRIA AND CHLOROPLASTS CONTAIN THEIR VERY OWN DNA.

ANYTHING ELSE?

I dunno... uh...maybe the mitochondria and chloroplasts are the same general size as the bacteria?

INDEED. **AND** THEY HAVE A **DOUBLE** MEMBRANE.

Why would they have two...?

Wait!

The inner membrane is the bacteria's original one, and the outer membrane is part of the eukaryote cell that engulfed it.

EXCELLENT! HE HAS A REMARKABLE MIND, YOUR HIGHNESS.

It's genetic.

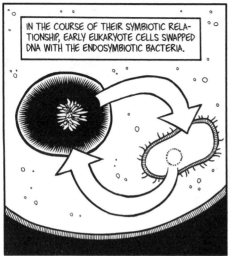

IN THE COURSE OF THEIR SYMBIOTIC RELATIONSHIP, EARLY EUKARYOTE CELLS SWAPPED DNA WITH THE ENDOSYMBIOTIC BACTERIA.

NOW THAT EACH HAD A BIT OF THE OTHER, NEITHER COULD FUNCTION PROPERLY ALONE.

YOU **NEED ME**.

PFFT. YOU NEED ME, TOO.

You say these eukaryotes evolved on Earth about 1.5 billion years ago, Bloort?

YES, YOUR GRAND PERCEPTIVENESS.

So, 3 billion years after life first formed on Earth it was **STILL** microscopic?

YES, BUT DON'T WORRY, YOUR HIGHNESS -- THESE EUKARYOTES ARE ABOUT TO KICK EVOLUTION INTO **HIGH GEAR**.

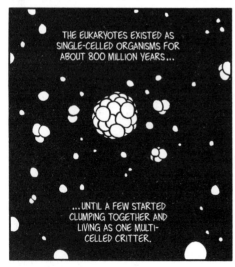

THE EUKARYOTES EXISTED AS SINGLE-CELLED ORGANISMS FOR ABOUT 800 MILLION YEARS...

...UNTIL A FEW STARTED CLUMPING TOGETHER AND LIVING AS ONE MULTI-CELLED CRITTER.

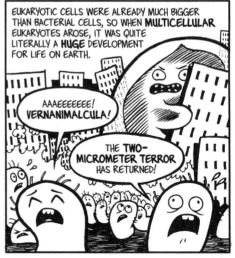

EUKARYOTIC CELLS WERE ALREADY MUCH BIGGER THAN BACTERIAL CELLS, SO WHEN **MULTICELLULAR** EUKARYOTES AROSE, IT WAS QUITE LITERALLY A **HUGE** DEVELOPMENT FOR LIFE ON EARTH.

AAAEEEEEEE! *VERNANIMALCULA!*

THE **TWO-MICROMETER TERROR** HAS RETURNED!

SWAPPING GENES

LATERAL TRANSFERS AND THE MODERN BDELLOIDIAN

BDELLOIDS [THE "B" IS SILENT] LIVE INTERESTING LIVES. THEY ARE MICROSCOPIC ANIMALS THAT RESIDE IN WATERY ENVIRONMENTS THAT FREQUENTLY DRY UP. FOR MOST CREATURES, THIS LOSS OF HABITAT WOULD SPELL **CERTAIN DEATH**, BUT THE BDELLOIDS JUST DRY UP INTO LITTLE BALLS AND WAIT. WHEN THE WATER RETURNS, THEY REHYDRATE AND SPRING BACK TO LIFE. SOME ARE CAPABLE OF SURVIVING IN THIS DRIED-OUT STATE FOR UP TO NINE YEARS.

BUT, WHEN THESE SLEEPING BEAUTIES AWAKE, THEY DON'T GO LOOKING FOR PRINCE CHARMINGS. THEY DON'T NEED THEM. BDELLOIDS ARE ALL FEMALE AND HAVE BEEN REPRODUCING **ASEXUALLY** FOR HUNDREDS OF MILLIONS OF YEARS. THIS HAS MADE THEM A BIT OF AN EVOLUTIONARY MYSTERY. MOST MULTICELLULAR ORGANISMS THAT HAVE COMPLETELY GIVEN UP SEX HAVE EVENTUALLY GIVEN UP THEIR EXISTENCE AS WELL. WITHOUT THE GENE-SHUFFLING BENEFITS OF SEXUAL REPRODUCTION, THEY TEND TO LOSE THE GENETIC VARIATION NECESSARY TO RESPOND TO RAPID ENVIRONMENTAL CHANGES AND RUN AN INCREASED RISK OF EXTINCTION.

SO HOW DO BDELLOIDIANS DO IT? RECENT WORK BY EUGENE GLADYSHEV, MATTHEW MESELSON, AND IRINA R. ARKHIPOVA HAS POINTED BIOLOGISTS TOWARD AN ANSWER. WHEN THE RESEARCHERS LOOKED AT THE GENES IN BDELLOIDS, THEY FOUND BITS OF DNA FROM PLANTS, FUNGI, AND OTHER BDELLOIDIANS. HOW DID THEY GET THERE WITHOUT SEXUAL REPRODUCTION? THE SCIENTISTS THINK IT HAPPENS WHEN THE BDELLOIDIANS WAKE UP.

ALTHOUGH THEY CAN SURVIVE BEING DRIED OUT, IT DOES TAKE ITS TOLL ON THE LITTLE CRITTERS. UPON REHYDRATION, ALL BDELLOIDIANS HAVE TO REBUILD MANY OF THEIR CELLS AND TIDY UP THEIR FRAGMENTED DNA. AS THEY'RE DOING REPAIRS, THEY TEND TO PICK UP MISCELLANEOUS SCRAPS OF DNA FLOATING IN THE WATER AND INCORPORATE THEM INTO THEIR GENOME. SOME OF THOSE SCRAPS COME FROM BDELLOIDIANS THAT RUPTURED AND DIED DURING THE PREVIOUS DRY-OUT.

SO, IT TURNS OUT THAT BDELLOIDS AREN'T COMPLETELY ASEXUAL AFTER ALL. THESE PERIODIC LATERAL TRANSFERS OF GENETIC MATERIAL HAVE MADE IT POSSIBLE FOR BDELLOIDIANS TO SURVIVE AND THRIVE FOR MILLIONS OF YEARS.

GETTING BIG PROVIDED SOME KEY EVOLUTIONARY ADVANTAGES. A SINGLE-CELLED BACTERIA HAS TO DO IT ALL: THAT ONE CELL MUST FIND FOOD, CONSUME AND DIGEST IT, BE IN CHARGE OF LOCOMOTION, RESPOND TO ITS ENVIRONMENT, AND DUMP WASTE.

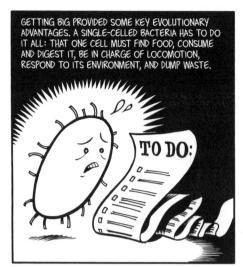

BUT MULTICELLULAR ORGANISMS HAVE ENOUGH CELLS TO SPLIT UP THE WORK. WHOLE GROUPS OF CELLS TAKE ON SPECIALIZED ROLES TO HELP ORGANISMS DEAL WITH DIFFERENT ENVIRONMENTAL STRESSES MORE EFFECTIVELY. THESE COHESIVE GROUPS OF CELLS WERE THE FIRST ORGANS.

HEY, GUYS, WHO WANTS TO PONDER SOMETHING?

CAN WE MULL IT OVER FIRST?

MMM, LET ME THINK ABOUT IT.

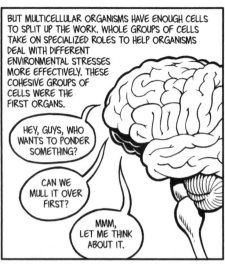

THE MULTICELLULAR EUKARYOTES WOULD SPAWN THREE MAJOR GROUPS OF ORGANISMS, AS WE SAW AT THE BEGINNING OF THE TOUR:

FUNGI, PLANTS, AND ANIMALS.

ALL THREE GROUPS HAVE MITOCHONDRIA IN THEIR CELLS, BUT ONLY PLANTS HAVE THE SUGAR-PRODUCING CHLOROPLASTS. CONSEQUENTLY, ALL OTHER MULTICELLULAR ORGANISMS RELY ON PLANTS, EITHER DIRECTLY OF INDIRECTLY, FOR THEIR STORED-UP SOLAR ENERGY.

STOP THAT! I'M MAKIN' **FOOD** HERE.

FUNGI MAKE THEIR LIVING **IN** OR **ON** OTHER ORGANISMS. MOSTLY, THEY DECOMPOSE DEAD THINGS. BUT A FEW ARE **PARASITES** THAT PREY ON THE LIVING...

...SOMETIMES EVEN TAKING CONTROL OF AN ANIMAL'S NERVOUS SYSTEM AND ALTERING ITS BEHAVIOR.

I THINK THERE'S A **FUNGUS** AMONG US.

MMRR...MUST... DISPERSE...FUNGUS... SPORES...

THE **ANIMALS** ARE THE THIRD GROUP OF MULTI-CELLULAR EUKARYOTES. THEY DISTINGUISH THEM-SELVES FROM FUNGI AND PLANTS IN TWO DRAMATIC WAYS. THEY **EAT OTHER ORGANISMS**, AND THEY **MOVE AROUND** A LOT.

YOU GUYS WANNA GO GET SOMETHING TO EAT?

HARDY-DEE-HAR-HAR.

REAL FUNNY, WISE GUY.

THE FIRST ANIMALS APPEARED ABOUT 560 MILLION YEARS AGO ON EARTH. THESE WERE SMALL AND SOFT-BODIED, AND THE MAJORITY WERE ROUND.

MANY RESEMBLED THE JELLYFISH, SEA ANEMONES, AND CORALS WE SEE TODAY.

BUT FOSSILS OF POSSIBLE WORM BURROWS FROM THIS TIME SUGGEST THAT A FEW SPECIES WERE SHAPED LIKE TUBES. THIS IS IMPORTANT BECAUSE ROUND ORGANISMS DON'T HAVE A FRONT END, AND ON EARTH, NO FRONT END MEANS NO BRAIN.

...

BUT TUBE-SHAPED ORGANISMS **DO** HAVE A FRONT END. AND, THE THING ABOUT A FRONT END -- OR A "HEAD," AS THEY CALL IT -- IS THAT IT IS THE PART OF THE ANIMAL THAT ENTERS AN ENVIRONMENT FIRST.

PARDON ME.

THINK NOTHING OF IT.

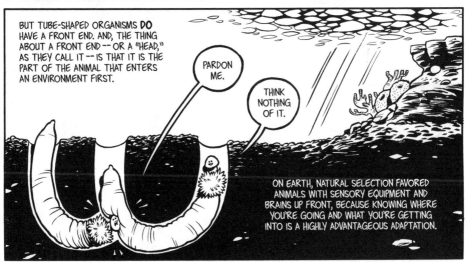

ON EARTH, NATURAL SELECTION FAVORED ANIMALS WITH SENSORY EQUIPMENT AND BRAINS UP FRONT, BECAUSE KNOWING WHERE YOU'RE GOING AND WHAT YOU'RE GETTING INTO IS A HIGHLY ADVANTAGEOUS ADAPTATION.

Is this the "high gear" you were referring to, Bloort?

COMING RIGHT UP, YOUR MAGNIFICENCE. IN FACT, IF THE PRINCE WOULD ASSIST ME...?

Sure thing, Bloort. What do I do?

I JUST NEED YOU TO PUSH THIS BIG RED BUTTON.

Okay.

But, what does "C.E." stand for?

C.E.

DO NOT TOUCH!

KA-BOOM

35

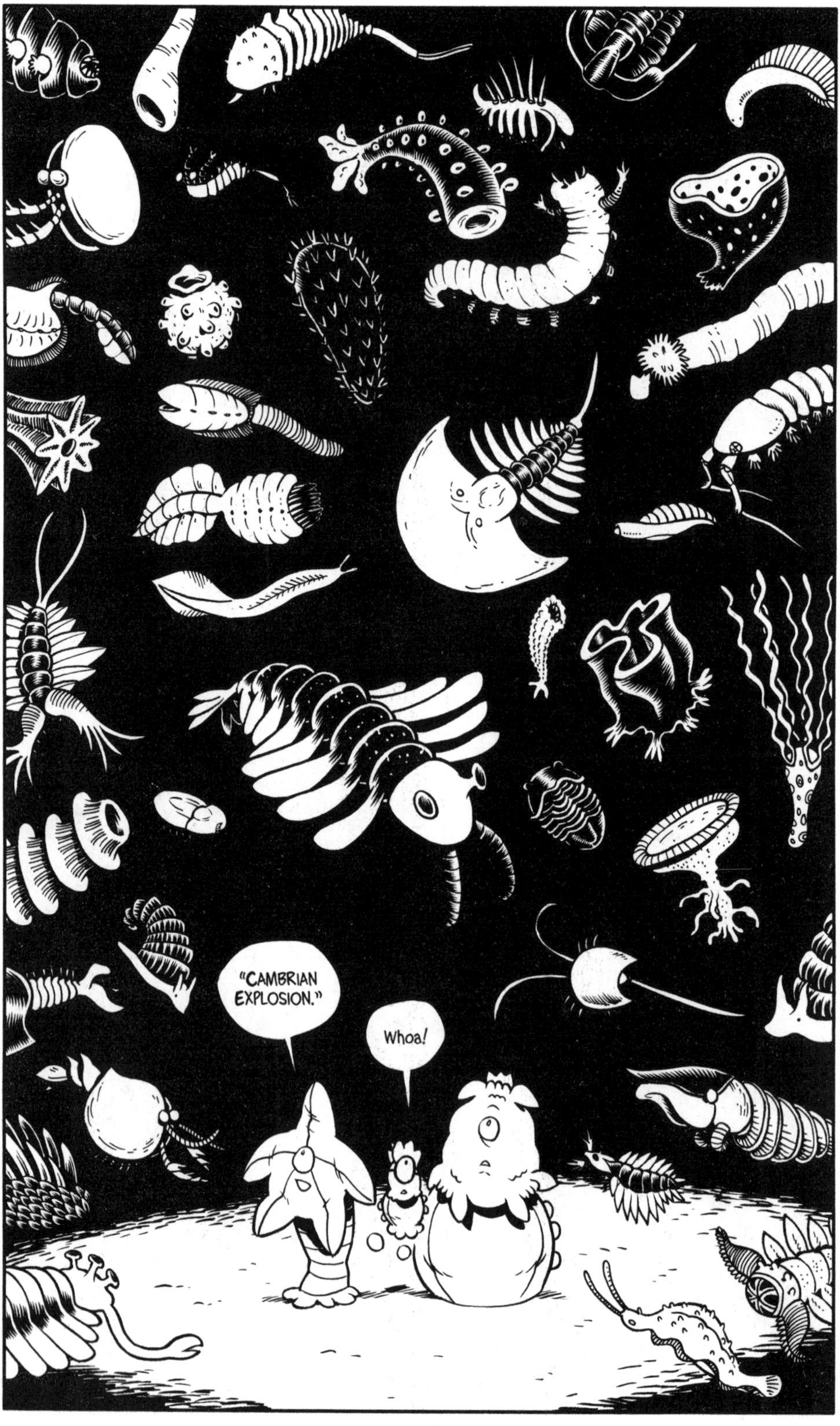
"CAMBRIAN EXPLOSION."
Whoa!

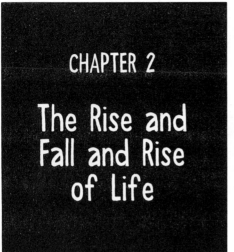

CHAPTER 2

The Rise and Fall and Rise of Life

These are amazing. They're so different from anything on Glargal.

THESE ARE THE PRODUCTS OF ONE OF EARTH'S FIRST GREAT EVOLUTIONARY EXPERIMENTS IN MULTICELLULAR ANIMAL LIFE.

What's **THIS** one?

OPABINIA. CHECK OUT ITS FIVE EYES AND THE GRASPING TENTACLE IN THE MIDDLE OF ITS FACE.

FIVE eyes?

Talk about **OVERKILL...**

OPABINIA

And what's this?

ANOMALOCARIS. IT WAS THREE FEET LONG -- ABOUT THE SIZE OF A THREE-YEAR-OLD HUMAN CHILD -- BUT IT WAS TEN TIMES BIGGER THAN ANY OTHER ANIMAL AT THE TIME, MAKING IT THE MONSTER PREDATOR OF EARTH'S OCEANS 500 MILLION YEARS AGO.

ANOMALOCARIS

Ah, I could tell as much from its regal bearing. Clearly this is the ancestor for these highly evolved humans we've heard so much about.

THE HUMANS AREN'T DESCENDED FROM ANOMALOCARIS, YOUR HIGHNESS. THEIR ANCESTOR IS OVER HERE.

PIKAIA.

What? That little sliver of a thing? Hmph. Not a very grand beginning, I must say.

Where did all these creatures come from, Bloort?

PIKAIA

WELL, YOUNG MAJESTY, THESE ARE **HOLOGRAMS**.

WHEN YOU PUSHED THAT RED BUTTON THE PROJECTORS WERE TURNED ON AND --

No, Bloort, I meant on **EARTH**.

Everything seems to have happened so fast. For four billion years Earth life was microscopic, and then **BOOM!** these large creatures appear out of nowhere.

THEIR SUDDEN APPEARANCE IN THE FOSSIL RECORD IS EXACTLY WHY EARTH SCIENTISTS REFERRED TO THEIR EMERGENCE AS THE **CAMBRIAN EXPLOSION**.

FOSSIL RECORD

Oh.

What's a **FOSSIL**?

Don't look at me. None of this was in his last report.

FOSSILS ARE ANY PRESERVED EVIDENCE OF ANCIENT LIFE. THEY CAN INCLUDE THE IMPRESSION OF A LEAF IN MUD, AN INSECT TRAPPED IN AMBER, PETRIFIED POOP, OR HARD ANIMAL PARTS LIKE SHELLS, EXOSKELETONS, AND BONES THAT HAVE TURNED TO ROCK.

Earth life can turn to stone? Their warriors must be **TERRIFYING**.

WELL, THEY TURN TO STONE ONLY AFTER THEY'RE **DEAD**.

They're **ZOMBIE** stone warriors? **AWESOME!**

NO, NO, NO, THEY'RE JUST **DEAD.** AND NOT EVERYTHING FOSSILIZES, ONLY THOSE ORGANISMS THAT DIE UNDER THE RIGHT CONDITIONS.

LET'S CONSIDER THIS DEAD TRILOBITE AS AN EXAMPLE.

MOST THINGS THAT DIE ARE QUICKLY GOBBLED UP BY BACTERIA AND FUNGI OR WORN AWAY BY ENVIRONMENTAL FORCES LIKE WIND AND RAIN.

TO HAVE A CHANCE OF BEING TURNED INTO A FOSSIL, AN ORGANISM USUALLY HAS TO DIE IN OR NEAR WATER AND THEN BE QUICKLY COVERED BY MUD.

THE SQUISHY PARTS OF THE TRILOBITE WILL DECAY NO MATTER WHAT YOU DO. BUT IF THE BODY IS BURIED, IT MIGHT BE PROTECTED LONG ENOUGH FROM THE FORCES OF DECAY FOR SOMETHING REALLY COOL TO HAPPEN.

YOU MAY SLOW US DOWN, BUT YOU CAN'T STOP US.

BACTERIA

MINERALS FROM THE SURROUNDING MUD START TO REPLACE THE MATERIAL THAT MAKES UP THE HARD PARTS OF THE ANIMALS. IT IS A SLOW PROCESS THAT PROCEEDS MOLECULE BY MOLECULE.

MUD

SILICA MOLECULES

TRILOBITE EXOSKELETON

EVENTUALLY, THE MINERALS COMPLETELY REPLACE THE BIOLOGICAL MATTER, AND THE ANIMAL'S BONES OR ARMOR HAVE BECOME STONE. THE ORGANISM IS NOW A FOSSIL.

AMAZING.

OVER MILLIONS OF YEARS, THESE LAYERS OF MUD CONTAINING FOSSILS ACCUMULATE -- ONE ATOP THE OTHER -- AND THROUGH THE FORCES OF WEIGHT AND COMPRESSION BECOME THE ROCKY LAYERS THAT FORM THE EARTH'S CRUST. SCIENTISTS CALL THESE LAYERS **STRATA.**

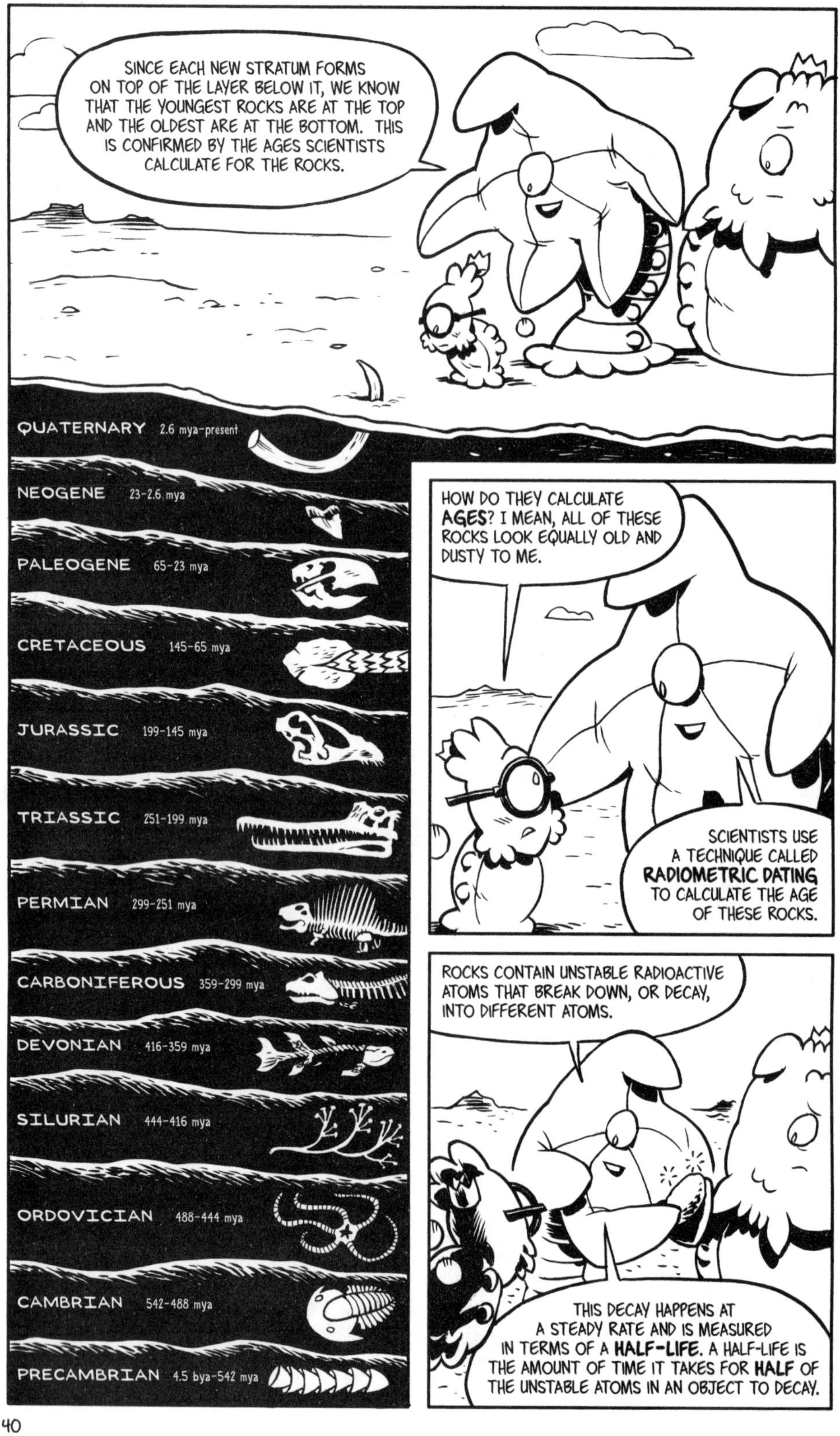
SINCE EACH NEW STRATUM FORMS ON TOP OF THE LAYER BELOW IT, WE KNOW THAT THE YOUNGEST ROCKS ARE AT THE TOP AND THE OLDEST ARE AT THE BOTTOM. THIS IS CONFIRMED BY THE AGES SCIENTISTS CALCULATE FOR THE ROCKS.
QUATERNARY 2.6 mya-present
NEOGENE 23-2.6 mya
PALEOGENE 65-23 mya
CRETACEOUS 145-65 mya
JURASSIC 199-145 mya
TRIASSIC 251-199 mya
PERMIAN 299-251 mya
CARBONIFEROUS 359-299 mya
DEVONIAN 416-359 mya
SILURIAN 444-416 mya
ORDOVICIAN 488-444 mya
CAMBRIAN 542-488 mya
PRECAMBRIAN 4.5 bya-542 mya
HOW DO THEY CALCULATE **AGES**? I MEAN, ALL OF THESE ROCKS LOOK EQUALLY OLD AND DUSTY TO ME.
SCIENTISTS USE A TECHNIQUE CALLED **RADIOMETRIC DATING** TO CALCULATE THE AGE OF THESE ROCKS.
ROCKS CONTAIN UNSTABLE RADIOACTIVE ATOMS THAT BREAK DOWN, OR DECAY, INTO DIFFERENT ATOMS.
THIS DECAY HAPPENS AT A STEADY RATE AND IS MEASURED IN TERMS OF A **HALF-LIFE**. A HALF-LIFE IS THE AMOUNT OF TIME IT TAKES FOR **HALF** OF THE UNSTABLE ATOMS IN AN OBJECT TO DECAY.

EARTH SCIENTISTS CAN MEASURE THE RATE OF DECAY TO CALCULATE HOW OLD A ROCK IS.

An example would be useful, Bloort.

OF COURSE, SIRE. HOW ABOUT THE DECAY OF **URANIUM-235** INTO **LEAD-206**?

 DECAY

URANIUM-235 (^{235}U)

LEAD-206 (^{206}Pb)

URANIUM-235 HAS A HALF-LIFE OF 704 MILLION YEARS.

LET'S IMAGINE WE HAVE A ROCK THAT CONTAINED EIGHT URANIUM-235 ATOMS WHEN IT FORMED ON THE ANCIENT EARTH. I'M USING A SMALL NUMBER HERE FOR ILLUSTRATION PURPOSES, YOUR HIGHNESS. THERE WOULD ACTUALLY BE MORE LIKE **8,000,000,000,000,000,000,000,000** ATOMS PRESENT!

Understood. Please proceed.

STARTING AT THE MOMENT THE ROCK FORMED, THE EIGHT URANIUM ATOMS BEGIN TO DECAY. AFTER THE FIRST 704 MILLION YEARS, FOUR OF THEM HAVE DECAYED INTO LEAD-206.

704 Million Years Later

U-RANIUM-235, ME LEAD-206

UGH. THAT'S THE BEST YOU COULD COME UP WITH AFTER 704 MILLION YEARS?

AFTER ANOTHER 704 MILLION YEARS, HALF OF THE REMAINING FOUR URANIUM-235 ATOMS WILL TURN INTO LEAD. NOW THREE-QUARTERS OF THE ATOMS ARE LEAD AND ONLY ONE-QUARTER ARE URANIUM.

1,408 Million Years

HEY, YOU GUYS, GET THE **LEAD** OUT!

I DON'T KNOW WHICH IS WORSE, THEIR JOKES OR THE FACT THAT WE'RE GONNA BE JUST LIKE THEM SOON.

And after another 704 million years there will be one Uranium atom and seven Lead atoms.

Ah, I see. By counting the relative number of the two atoms, scientists can calculate the age of these rocks.

PRECISELY. OF COURSE, SCIENTISTS ACTUALLY TAKE MULTIPLE READINGS USING DIFFERENT TYPES OF ATOMS WITH DIFFERENT HALF-LIVES TO CONFIRM THEIR CALCULATIONS.

THE STONE GRIFFINS OF SAKA-SCYTHIA

DRAGONS, CYCLOPSES, AND GRIFFINS ARE ALL MYTHOLOGICAL CREATURES THAT HAVE NO BASIS IN REALITY. OR DO THEY? NEW EVIDENCE SUGGESTS THAT SOME OF THESE CREATURES WERE MORE THAN JUST FANCIFUL STORIES. THANKS TO THE WRITINGS OF ANCIENT GREEK HISTORIANS, IT NOW APPEARS THAT FOSSILS WERE COMMONLY KNOWN TO PEOPLE LIVING IN THE MIDDLE EAST AND ASIA AS MANY AS 4,500 YEARS AGO.

BACK THEN, FARMERS BELIEVED THAT FOSSIL BONES THEY UNEARTHED IN THEIR FIELDS BELONGED TO **PRE-HISTORIC DRAGONS**. LIKEWISE, MAMMOTH SKULLS FOUND IN CAVES STREWN WITH LARGE BONES MAY HAVE SPAWNED THE LEGEND OF THE **CYCLOPS**. CYCLOPSES WERE BELIEVED TO BE A RACE OF ONE-EYED, MAN-EATING GIANTS. MORE LIKELY, THESE WERE THE BONES AND SKULLS OF AN-CIENT MAMMOTHS, WHICH WERE ELEPHANT-LIKE ANIMALS. THE SKULL OF A MAMMOTH HAS A LARGE CENTRAL HOLE AT BROW LEVEL, THROUGH WHICH THE ANIMAL'S TRUNK DESCENDS. IT CAN EASILY BE MISTAKEN FOR A LARGE, SINGLE EYEHOLE.

RESEARCHER ADRIENNE MAYOR WONDERED IF FOSSIL REMAINS MAY HAVE ALSO INSPIRED THE STORY OF THE **GRIFFIN**. GRIFFINS ARE MYTHOLOGICAL BEASTS WITH THE HEAD AND BEAK OF AN EAGLE AND THE BODY OF A LION. NOMADIC TRIBES IN THE MOUNTAINS OF SAKA-SCYTHIA, IN MODERN-DAY UZBEKISTAN, STILL TELL TALES OF THE FIERCE GRIFFINS WHO ROAMED THE DESERTS AND LINED THEIR NESTS WITH GOLD. COULD THESE BEASTS HAVE A FOOT (OR A CLAW) IN REALITY? MAYOR TRACED THE STORIES TO THEIR SOURCE AND MADE AN EXCITING DISCOVERY.

IN THE MOUNTAINS OF SAKA-SCYTHIA, PALEON-TOLOGISTS FOUND A TREASURE TROVE OF FOSSIL BONES AND EGGS BELONGING TO A DINOSAUR NAMED **PROTOCERATOPS**. MOST OF THESE PROTOCERATOPS SKELETONS ARE COM-PLETELY INTACT AND LOOK SURPRISINGLY FAMILIAR; THEY HAVE A LARGE, BIRDLIKE BEAK AND ARE ABOUT THE SIZE OF A LION. NOMADS TRAVELING THE AREA HAD UNDOUBTEDLY SEEN PROTOCERATOPS FOSSILS EXPOSED BY THE EVER-SHIFTING SANDS. MAYOR SUGGESTS THAT THE LEGEND OF THE GRIFFIN MAY HAVE BEEN INSPIRED BY THESE MAGNIFICENT ANIMALS OF STONE.

DID THE PROTOCERATOPS ALSO LINE THEIR NESTS WITH GOLD, AS THE MYTHOLOGICAL GRIFFINS WERE SAID TO DO? PROBABLY NOT. BUT THE MOUNTAINS OF THE REGION DO HAVE RICH GOLD DEPOSITS THAT ERODE TO FORM GOLD DUST, WHICH CAN OFTEN BE FOUND SPRINKLED ACROSS THE DESERT AND IN THE FOSSILIZED NESTS OF DINOSAURS.

THE LOWEST LAYER -- OR STRATUM -- HERE WAS FORMED DURING THE 4 BILLION YEARS OF THE PRECAMBRIAN ERA. THE FIRST EARTH GEOLOGISTS TO EXAMINE IT WERE STRUCK BY THE FACT THAT IT **APPEARED** TO LACK FOSSILS COMPLETELY.

ORDOVICIAN

CAMBRIAN

PRECAMBRIAN

But you told us there **WERE** critters in the Precambrian. How could you know that if there were no fossils?

OH, THERE ARE FOSSILS IN THE ROCK. THEY'RE JUST REALLY SMALL AND HARD TO SPOT. RE-SEARCHERS HAVE NOW FOUND LOTS OF JELLYFISH AND SPONGES AND TEN SAMPLES OF VERNANIMICULA, ONE OF THE FIRST CRITTERS WITH A FRONT END.

THE FOSSILS WERE JUST TOO TINY TO SPOT WITHOUT MODERN MICROSCOPES. WHEN EARLY SCIENTISTS COM-PARED THE PRECAMBRIAN LAYER TO THE FOSSIL-RICH LAYER RIGHT ABOVE IT, THEY THOUGHT THEY'D DISCOVERED A SUDDEN TRANSITION FROM NO LIVING THINGS TO AN ABUNDANCE OF ORGANISMS.

Ahh -- so life seemed to "explode" during the Cambrian.

CAMBRIAN

PRECAMBRIAN

But how or why did those **SMALL SQUISHY THINGS** of the Precambrian turn into the **BIG HARD CREATURES** of the Cambrian?

THAT IS PRECISELY THE QUESTION, MY PRINCE.

SCIENTISTS THINK THEY PROBABLY GOT BIG BECAUSE **OXYGEN LEVELS** IN THE ATMO-SPHERE CONTINUED TO **RISE** AS A RESULT OF PHOTOSYNTHESIS.

AN

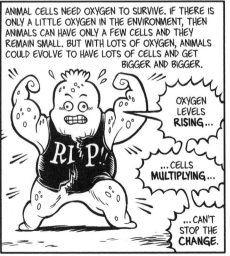

ANIMAL CELLS NEED OXYGEN TO SURVIVE. IF THERE IS ONLY A LITTLE OXYGEN IN THE ENVIRONMENT, THEN ANIMALS CAN HAVE ONLY A FEW CELLS AND THEY REMAIN SMALL. BUT WITH LOTS OF OXYGEN, ANIMALS COULD EVOLVE TO HAVE LOTS OF CELLS AND GET BIGGER AND BIGGER.

OXYGEN LEVELS **RISING**...

R.I.P!

...CELLS **MULTIPLYING**...

...CAN'T STOP THE **CHANGE.**

Just like an abundance of food can allow some squinches to get bigger and bigger?

AHEM.

WELL, I...UM.

Do not mock my regal magnitude in front of Bloort, son.

43

WHERE WAS I? OH, YES, INCREASED OXYGEN LEVELS...

That might explain why they got big, Bloort. But it doesn't explain the sudden appearance of hard parts that fossilize well, does it?

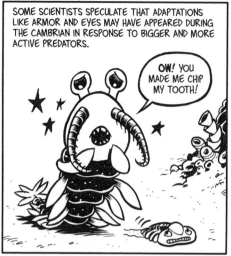

SOME SCIENTISTS SPECULATE THAT ADAPTATIONS LIKE ARMOR AND EYES MAY HAVE APPEARED DURING THE CAMBRIAN IN RESPONSE TO BIGGER AND MORE ACTIVE PREDATORS.

OW! YOU MADE ME CHIP MY TOOTH!

WHATEVER THE CAUSE, LIFE CHANGED DRAMATICALLY IN THE CAMBRIAN AND BEGAN AN ENORMOUS RADIATION INTO THE OCEAN.

RADIATION? They can turn to stone AND they have radioactive powers? Just like SUPER-DUPER SQUINCH?

Sigh, those comics are going to rot his brain.

ACTUALLY, YOUNG MAJESTY, I'M TALKING ABOUT AN EVOLUTIONARY RADIATION. THIS IS A RAPID INCREASE IN THE NUMBER OF SPECIES AS A RESULT OF NEW ADAPTATIONS.

RADIATIONS ARE TIMES OF EXUBERANT EXPERIMENTATION. AS ORGANISMS GOT BIGGER AND HAD MORE CELLS, LIFE EXPERIMENTED WITH NEW WAYS OF GETTING RESOURCES.

CAMBRIAN SPONGES LIKE THIS ARCHAEOCYATH SAT ANCHORED TO REEFS AND ATE THE BACTERIA AND SINGLE-CELL PLANTS -- CALLED PHYTOPLANKTON -- THAT IT FILTERED FROM THE WATER.

Y'KNOW, ALL THIS FILTER-FEEDING IS REALLY STRAINING.

CAMBRIAN JELLYFISH DINED ON TINY FLOATING ORGANISMS.

SMALL ANIMALS LIKE **WIWAXIA** GRAZED ON THE MATS OF CYANOBACTERIA ON THE OCEAN FLOOR, AND ANCIENT CLAMS BURIED THEMSELVES IN THE MUD AND SUCKED FOOD FROM THE WATER.

WIWAXIA

I'VE BEEN HIT! I'VE BEEN HIT.

PFFT, HE JUST **GRAZED** YOU.

PREDATORY WORMS BURROWED UNDER THE OCEAN FLOOR, HUNTING FOR FOOD, WHILE GIANTS LIKE ANOMALOCARIS CRUISED THE WATERS ABOVE, PREYING UPON THEM ALL.

!

AND WHEN CREATURES EVENTUALLY DIED, SCAVENGERS DEALT WITH THEIR REMAINS.

So, for 4 billion years life on Earth was microscopic. Then, about a half billion years ago, increased oxygen levels allowed animals to get bigger.

The increased oxygen was a by-product of cyanobacteria and phytoplankton.

All of this allowed early life forms to fill the ocean with many new species within a few million years.

Bigger animals could take on specialized forms and play specialized roles in their environment.

What? Did I **MISS** something?

NO, IT'S JUST THAT I THINK YOU SAID THAT BETTER THAN I DID...YOUR MAJESTY.

You're just an excellent teacher, Bloort.

I HOPE YOUR **FATHER'S** LISTENING.

Um, I do have a meeting with the High Council later today, so if we could keep moving...?

OF COURSE, SIRE.

I APPRECIATE YOUR ROYAL INDULGENCE.

I THINK WE HAVE SUFFICIENT BACKGROUND NOW THAT WE CAN MOVE THROUGH THE NEXT SEVERAL HUNDRED MILLION YEARS AT A BRISK PACE.

THE FOSSIL RECORD OF THE CAMBRIAN PERIOD CONTAINS EXAMPLES OF VIRTUALLY EVERY MODERN ANIMAL FORM ON EARTH. OF PARTICULAR INTEREST IS THE APPEARANCE OF **CHORDATES** AND **ARTHROPODS**.

CHORDATES ARE ANIMALS THAT HAVE A THIN, FLEXIBLE ROD CALLED A **NOTOCHORD** RUNNING DOWN THEIR BACKS AND A **NERVE CORD** RIGHT ABOVE THE NOTOCHORD. THE **PIKAIA** WE SAW EARLIER IS AN EXAMPLE OF A CHORDATE.

PIKAIA

It is incredible to believe that this creature's descendants will one day write poetry and untangle the mysteries of DNA.

LATE IN THE CAMBRIAN PERIOD, THE NOTOCHORDS OF ONE GROUP OF CHORDATES WOULD EVOLVE INTO A RIGID INTERNAL SUPPORT STRUCTURE CALLED A **BACKBONE**. ORGANISMS WITH THIS ADAPTATION ARE CALLED **VERTEBRATES** AND ARE THE GROUP TO WHICH HUMAN BEINGS BELONG.

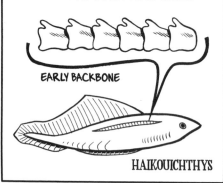

EARLY BACKBONE

HAIKOUICHTHYS

ARTHROPODS, ON THE OTHER HAND, HAVE THEIR RIGID SUPPORT STRUCTURE ON THE OUTSIDE. THEY'RE COVERED IN A HARD SHELL CALLED AN **EXOSKELETON**. AND THEY ALSO HAVE JOINTED LEGS AND BODIES MADE UP OF MULTIPLE SEGMENTS. MODERN ARTHROPODS INCLUDE ORGANISMS SUCH AS CRABS, CENTIPEDES, SPIDERS, AND INSECTS.

C'MON, CUT ME SOME SLACK HERE. WE'RE **RELATED**, FOR PETE'S SAKE!

EXOSKELETON

IN THE CAMBRIAN, THE DOMINANT ARTHROPOD WAS THE **TRILOBITE**. THEY WEREN'T THE BIGGEST ANIMALS -- USUALLY ONLY ONE TO FOUR INCHES IN LENGTH -- BUT THERE WERE A **LOT** OF THEM AND THEY EVOLVED INTO A WIDE VARIETY OF PREDATORS, SCAVENGERS, AND EVEN FILTER-FEEDERS.

WE CAN DO IT **ALL**, BABY.

PARTY!

PARTY!

MEANWHILE, ON LAND, NOT MUCH WAS HAPPENING.

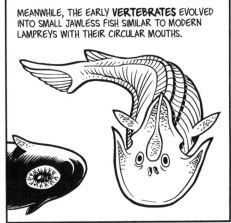

IN THE ORDOVICIAN PERIOD, TRILOBITES REACHED THE PEAK OF THEIR SUCCESS IN THE ANCIENT OCEANS.

WE ARE UNSTOPPABLE!

TRILOBITES RULE!

MEANWHILE, THE EARLY **VERTEBRATES** EVOLVED INTO SMALL JAWLESS FISH SIMILAR TO MODERN LAMPREYS WITH THEIR CIRCULAR MOUTHS.

THIS WAS ALSO WHEN **ECHINODERMS**, THE GROUP OF ORGANISMS CONTAINING SEA STARS AND SEA CUCUMBERS, WENT THROUGH A SUBSTANTIAL RADIATION.

GO, BABY, GO!

YOU CAN DO IT!

ON LAND, THE FIRST TINY PLANTS APPEARED. THEY WERE SMALL AND MOSSY AND NEEDED TO STAY CLOSE TO THE WATER.

DON'T LEAVE ME.

I WON'T GO FAR.

FINALLY, THE FIRST FUNGI OOZED ONTO LAND DURING THE ORDOVICIAN AND PROMPTLY BEGAN DECOMPOSING DEAD PLANTSTUFF.

I'M NOT WHAT YOU WOULD CALL A PICKY EATER.

DURING THE SILURIAN PERIOD, THE NUMBER OF TRILOBITE SPECIES DECLINED SUBSTANTIALLY.

THAT'S FINE. I NEEDED A LITTLE "ME" TIME.

What do you mean, they "declined"? They're at the top of their game. Look at them. They're everywhere.

Plus, they're really cute.

THE CAUSE OF THEIR DECLINE ISN'T CLEAR, YOUR HIGH-NESSES.

BUT THEIR DROP IN NUMBERS DOES COINCIDE WITH THE EVOLUTION OF SHARKS AND BONY FISH WITH JAWS. PERHAPS TRILOBITES WERE AN ABUNDANT SOURCE OF SLOW-MOVING FOOD?

Shoo, villains!

Scat!

ELSEWHERE IN THE SEA, THE FIRST CORALS AND SEAWEED APPEARED.

WHILE ON LAND, THE FIRST VASCULAR PLANTS WERE EVOLVING. THESE PLANTS HAD RIGID TUBES IN THEIR STEMS THAT HELPED THEM STAND UP TALL AND SUCK UP WATER FROM THE GROUND.

THINGS ARE LOOKING UP.

THIS WAS AN ENORMOUSLY IMPORTANT ADAPTATION FOR LIVING ON LAND AND WOULD ONE DAY MAKE IT POSSIBLE FOR REDWOOD TREES TO STAND MORE THAN 300 FEET TALL.

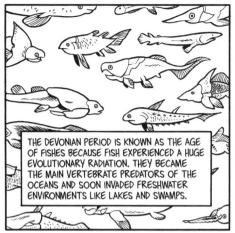

THE DEVONIAN PERIOD IS KNOWN AS THE AGE OF FISHES BECAUSE FISH EXPERIENCED A HUGE EVOLUTIONARY RADIATION. THEY BECAME THE MAIN VERTEBRATE PREDATORS OF THE OCEANS AND SOON INVADED FRESHWATER ENVIRONMENTS LIKE LAKES AND SWAMPS.

MEANWHILE, TRILOBITE NUMBERS CONTINUED TO DECLINE.

WE KEEP **LOSING** SPECIES.

IT'S TIME WE STARTED A **BUDDY SYSTEM.** HOLD MY HAND.

MMM...TWO FOR THE PRICE OF ONE!

ON LAND, FERNS AND VASCULAR PLANT SPECIES RADIATED INTO NEW ENVIRONMENTS AND THE **FIRST INSECTS** EVOLVED FROM SMALL MARINE ARTHROPODS THAT HAD CRAWLED ONTO LAND SOMETIME IN THE SILURIAN.

BY THE END OF THE DEVONIAN'S 50-MILLION-YEAR SPAN, PLANTS HAD EVOLVED **SEEDS,** INSECTS HAD EVOLVED **WINGS,** AND FRESHWATER FISH SPAWNED THE NEXT MAJOR GROUP OF VERTEBRATES: THE **AMPHIBIANS.**

LOOK AT ALL THAT **FOOD...**

AMPHIBIANS ARE VERTEBRATES THAT CAN LIVE ON LAND AND IN THE WATER. THE YOUNG BREATHE WATER LIKE A FISH, BUT THE ADULTS BREATHE AIR AND CAN VENTURE ONTO LAND.

BE CAREFUL, YOU'LL DRY OUT WITH THAT THIN SKIN OF YOURS.

DON'T WORRY.

AMPHIBIANS TOOK THEIR FIRST STEPS ONTO LAND USING STURDY LEGS THAT HAD EVOLVED FROM FRAGILE FISH FINS.

THIS ISN'T AS EASY AS IT LOOKS.

CARBONIFEROUS PERIOD 359–299 MILLION YRS AGO

BY THE TIME THE CARBONIFEROUS PERIOD ROLLED AROUND THERE WAS ONLY ONE GROUP OF TRILOBITES LEFT AND ONLY A HANDFUL OF SPECIES.

OH, DEAR.

FISH SHARED DOMINION OF THE SEAS WITH RELATIVES OF MODERN SQUID CALLED **AMMONOIDS**.

YOU TAKE THE TRILOBITE.

NO, **YOU**, I INSIST.

URP! NO, REALLY... I COULDN'T.

MUST... SCURRY... FASTER...

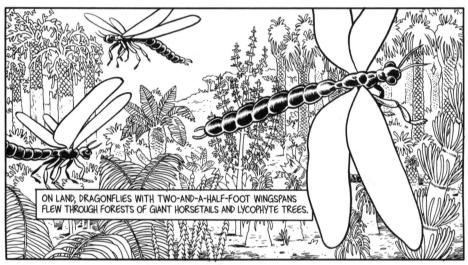

ON LAND, DRAGONFLIES WITH TWO-AND-A-HALF-FOOT WINGSPANS FLEW THROUGH FORESTS OF GIANT HORSETAILS AND LYCOPHYTE TREES.

THE AMPHIBIANS BEGAN TO RADIATE ON LAND AND IN THE WATER OF STREAMS, RIVERS, AND LAKES.

DURING THIS ERA, ONE GROUP OF AMPHIBIANS WOULD GIVE RISE TO A NEW GROUP OF VERTEBRATES CALLED **REPTILES** THAT HAD WATERTIGHT SKIN AND EGGS THAT COULD BE LAID AND SURVIVE ON LAND.

THE PERMIAN IS SOMETIMES REFERRED TO AS THE AGE OF AMPHIBIANS. DURING THIS TIME AMPHIBIANS AND REPTILES WERE THE DOMINANT VERTEBRATES IN FRESHWATER AND ON LAND.

HIGH-FIVE!!

LIFE IN THE PERMIAN OCEANS WAS MUCH LIKE IT HAD BEEN DURING THE CARBONIFEROUS. BUT ON LAND, PLANTS CALLED **GYMNOSPERMS** HAD EVOLVED AND WERE RAPIDLY SPREADING ACROSS THE LANDSCAPE. GYMNOSPERMS HAVE SEEDS ENCLOSED IN A CONE, LIKE MODERN-DAY PINE TREES.

BUT AS THE PERMIAN PERIOD DREW TO A CLOSE, THE FACE OF THE PLANET WAS SLOWLY CHANGING. THE CONTINENTS OF EARTH BEGAN TO COME TOGETHER TO FORM ONE GIGANTIC LAND MASS CALLED **PANGEA.**

The land **MOVES?**

YES, YOUR HIGHNESS. THE EARTH'S **CRUST** IS LIKE A JIGSAW PUZZLE COMPOSED OF MANY PIECES SITTING ON A LAYER OF LIQUID ROCK CALLED THE **MANTLE.**

So, these pieces of the Earth's crust are sorta **FLOATING** on the mantle?

YES. AND THEY MOVE VERY SLOWLY -- A LITTLE LESS THAN A HALF-INCH TO MAYBE A FULL INCH A YEAR. WHEN THEY FINALLY COLLIDED, THE SUPERCONTINENT OF PANGEA FORMED... AND SOMETHING STUNNING HAPPENED TO LIFE ON EARTH.

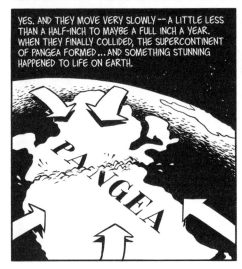

Let me guess: another **EXPLOSION?**

NOT QUITE, YOUR HIGHNESS. AT THE END OF THE PERMIAN...

P.E.

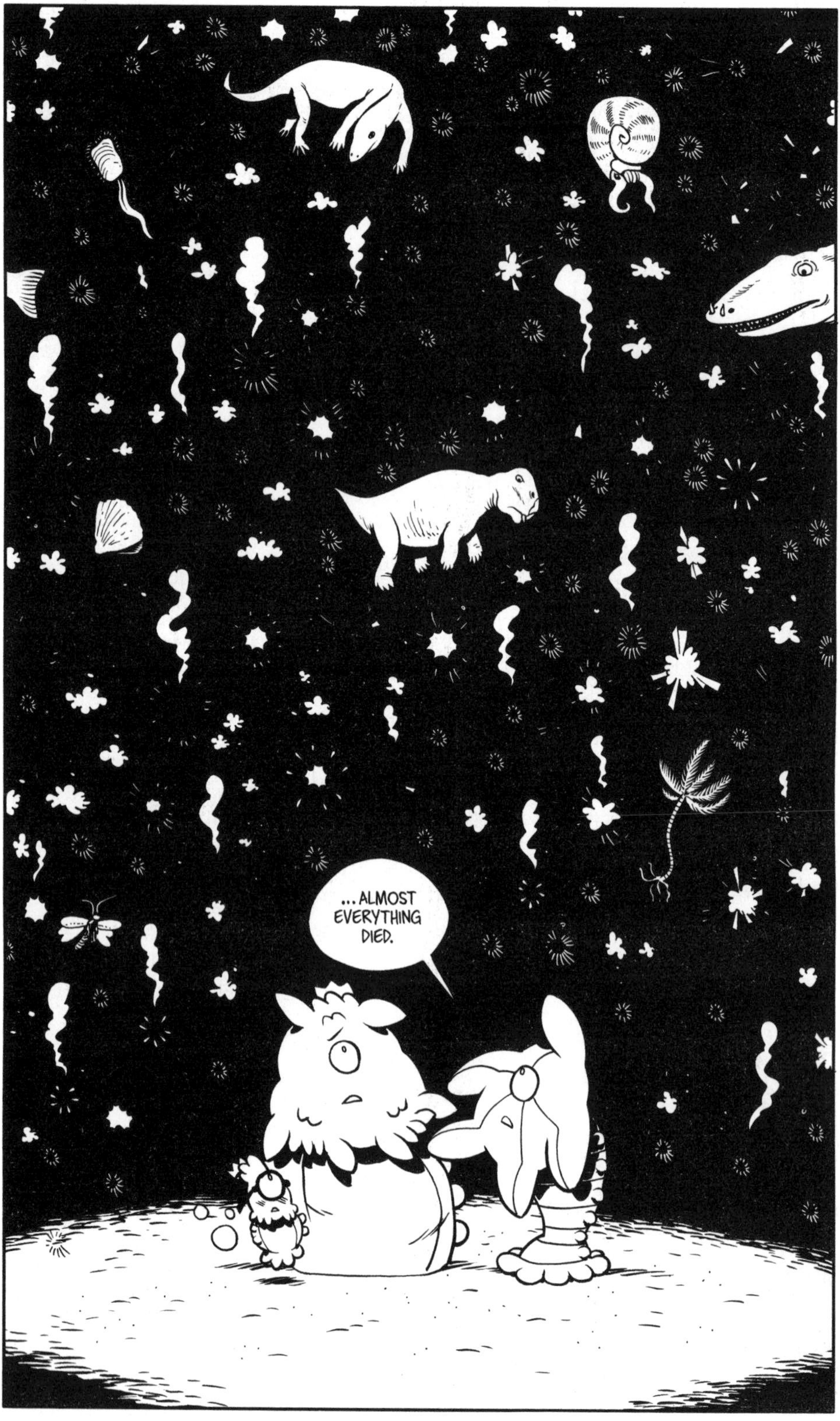
...ALMOST EVERYTHING DIED.

CHAPTER 3

E Is for Extinction

Great Glargal, what just happened?

LIFE ON EARTH WAS JUST DEALT A DEVASTATING BLOW.

AT THE END OF THE PERMIAN PERIOD 80 TO 90% OF SPECIES IN THE SEA WERE LOST, TWO-THIRDS OF THE AMPHIBIANS AND REPTILES DIED, ONE-THIRD OF THE INSECTS DISAPPEARED, AND THE FORESTS AROUND THE GLOBE WERE DEVASTATED.

I thought for sure the **P.E.** on that button you pushed stood for Permian **EXPLOSION**.

I'M AFRAID THE **E** IS FOR **EXTINCTION**, OH SUPREMELY COMPASSIONATE ONE.

Well, I don't know what extinction is, but I am fairly certain I don't like it.

EXTINCTION IS WHEN A SPECIES COMPLETELY DIES OUT, AND I'M AFRAID IT IS VERY MUCH A FACT OF LIFE. AND DEATH.

LITERALLY.

THE EXTINCTION EVENT AT THE END OF THE PERMIAN WAS THE BIGGEST **MASS EXTINCTION** IN EARTH'S HISTORY. MASS EXTINCTIONS INVOLVE THE DEATHS OF ENORMOUS NUMBERS OF SPECIES ALL OVER THE PLANET.

And the trilobites?

DIDN'T MAKE IT.

THERE HAVE BEEN FIVE MASS EXTINCTIONS IN EARTH'S HISTORY. TRILOBITES SURVIVED THE FIRST TWO IN THE ORDOVICIAN AND DEVONIAN, BUT THE PERMIAN EXTINCTION WAS MORE THAN THEY COULD HANDLE.

STOP THE WORLD, I'M GETTING OFF.

CAN WE DO THAT?

Entire species disappearing **FOREVER**? I find the whole notion terrifying, Bloort.

SO DID MANY EARTH SCIENTISTS WHEN GEORGES CUVIER FIRST PROPOSED THE CONCEPT OF EXTINCTION IN 1796.

I DID A DETAILED EXAMINATION OF THE BONES OF A WOOLLY MAMMOTH AND COMPARED THEM TO THE BONES OF AFRICAN AND INDIAN ELEPHANTS. IT WAS CLEAR TO ME THAT THEY WERE NOT THE SAME.

AFRICAN   **WOOLLY**   **INDIAN**

SO, AS THERE WERE NO WOOLLY MAMMOTHS STILL ROAMING ABOUT EUROPE, I PROPOSED THAT THE SPECIES WAS NO MORE. IN FACT, THIS WAS TRUE FOR MANY OF THE FOSSIL SPECIES I EXAMINED.

FROM THIS I FURTHER CONCLUDED, AND I QUOTE MYSELF: "ALL OF THESE FACTS, CONSISTENT AMONG THEMSELVES, AND NOT OPPOSED BY ANY REPORT, SEEM TO ME TO PROVE THE EXISTENCE OF A WORLD PREVIOUS TO OURS, DESTROYED BY SOME KIND OF CATASTROPHE."

THE CATASTROPHE CUVIER WAS DESCRIBING WAS **EXTINCTION**, AND IT FLEW IN THE FACE OF THE GENERALLY HELD BELIEF THAT SPECIES WERE PERFECT AND COULD NOT GO EXTINCT.

YOU ARE **PERFECT**, DON'T **EVER** CHANGE.

BEST FRIENDS FOREVER?

NO, **YOU'RE** PERFECT.

TOTALLY.

And yet, life must have found a way to go on after the Permian, right, Bloort?

ABSOLUTELY.

THE PERMIAN EXTINCTION WIPED MUCH OF THE SLATE CLEAN AND OPENED DOORS FOR MANY SPECIES LIVING IN THE MARGINS.

IN THE AFTERMATH OF A MASS EXTINCTION, THE WORLD CAN LOOK MUCH DIFFERENT FROM THE ONE BEFORE THE EXTINCTION EVENT. NEW SPECIES EVOLVE FROM THE SURVIVORS AND RADIATE INTO A WIDE-OPEN WORLD.

TAKE **INSECTS**, FOR EXAMPLE. REMEMBER THAT GIANT DRAGONFLY WE SAW IN THE CARBONIFEROUS?

HUFF PUFF

The one with the two-and-a-half-foot wingspan?

THE SAME. IT WAS A MEMBER OF THE FIRST GROUP OF INSECTS TO EVOLVE WINGS. THIS GROUP WAS CALLED THE **PALEOPTERA** -- WHICH MEANS "OLD WING" -- AND THEY'RE EASY TO SPOT BECAUSE THEY CAN'T FOLD THEIR WINGS BACK WHEN THEY ARE RESTING.

THE PALEOPTERA DOMINATED THE CARBONIFEROUS AND PERMIAN FORESTS, BUT THERE WAS ANOTHER GROUP OF INSECTS LIVING IN THEIR SHADOW CALLED THE **NEOPTERA** -- "NEW WING" -- THAT INCLUDED THE COCKROACHES AND GRASSHOPPERS.

SOME DAY I HOPE TO MAKE IT BIG JUST LIKE YOU.

ME, TOO!

AH, THE LITTLE PEOPLE.

THE NEOPTERA HAD EVOLVED A HINGE MECHANISM THAT MADE IT POSSIBLE FOR THEM TO FOLD THEIR WINGS BACK WHEN THEY WEREN'T IN USE.

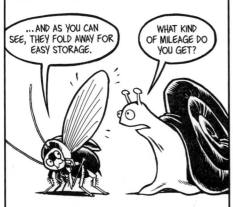

...AND AS YOU CAN SEE, THEY FOLD AWAY FOR EASY STORAGE.

WHAT KIND OF MILEAGE DO YOU GET?

THE WING HINGE ADAPTATION PROTECTED THEIR WINGS FROM GETTING BANGED UP AND ALLOWED THE NEOPTERA TO GET INTO SMALLER PLACES THE PALEOPTERA COULDN'T GO.

I'M SORRY, SIR, BUT WE CAN'T FIT YOU THROUGH THE DOOR.

NUTS.

Café NEO

With an adaptation like that they must have quickly replaced the paleoptera.

YOU MIGHT THINK SO, BUT THAT WASN'T THE CASE. BECAUSE THE PALEOPTERA HAD EVOLVED FIRST, THEY OCCUPIED MANY NICHES ON LAND BEFORE THE NEOPTERA CAME ON THE SCENE.

THE EXTINCTION AT THE END OF THE PERMIAN UPSET THAT BALANCE BY WIPING OUT ONE-THIRD OF ALL INSECT SPECIES. WHEN THE SURVIVORS BEGAN RADIATING DURING THE TRIASSIC, THE NEOPTERANS WON THE RACE. NOW, 98% OF ALL INSECT SPECIES ON EARTH CAN FOLD THEIR WINGS BACK.

What are the vertebrates doing at this time?

THEY'RE REBOUNDING AS WELL, BUT THE TRIASSIC PERIOD WOULD BE ESPECIALLY GOOD FOR THE REPTILES.

THE REPTILES RADIATED DRAMATICALLY IN THE TRIASSIC. FLYING REPTILES KNOWN AS PTEROSAURS EVOLVED DURING THE TRIASSIC AND WERE THE FIRST VERTEBRATES TO CONQUER THE AIR.

NOW IF I COULD ONLY CONQUER MY FEAR OF HEIGHTS!

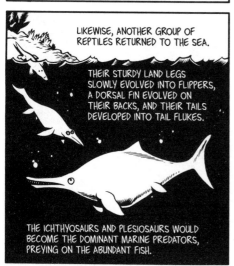

LIKEWISE, ANOTHER GROUP OF REPTILES RETURNED TO THE SEA.

THEIR STURDY LAND LEGS SLOWLY EVOLVED INTO FLIPPERS, A DORSAL FIN EVOLVED ON THEIR BACKS, AND THEIR TAILS DEVELOPED INTO TAIL FLUKES.

THE ICHTHYOSAURS AND PLESIOSAURS WOULD BECOME THE DOMINANT MARINE PREDATORS, PREYING ON THE ABUNDANT FISH.

Whoa, hold on a minute. Why would reptiles **RETURN TO THE WATER**? Didn't their ancestors just come from there?

EVOLUTION IS NOT A PROGRESSIVE MARCH, MOST SAGACIOUS EXCELLENCY.

LIFE HAS NO DESTINATION, NO ULTIMATE GOAL. IT EVOLVES TO TAKE ADVANTAGE OF NEW WAYS OF GETTING RESOURCES. IN THIS CASE, A SEA FULL OF FISH IS A LOVELY RESOURCE FOR A REPTILE THAT CAN SWIM.

I see.

That's the sea and air, what about the land, Bloort?

REPTILES ON THE LAND WOULD RADIATE AND GIVE RISE TO SEVERAL GROUPS. THESE INCLUDE THE SNAKES AND LIZARDS, THE CROCODILIANS, AND FIVE GROUPS OF GIANT REPTILES...

Five **DIFFERENT** groups of giant reptiles? Astonishing.

YES.

OF COURSE, ONLY ONE WOULD SURVIVE THE MASS EXTINCTION EVENT AT THE END OF THE TRIASSIC.

What? **ANOTHER** extinction? So soon?

THESE THINGS AREN'T **PLANNED**, SIRE.

JURASSIC & CRETACEOUS PERIODS

ABOUT HALF OF THE SPECIES LIVING ON EARTH WENT EXTINCT AT THE END OF THE TRIASSIC.

AS WE MOVE INTO THE JURASSIC AND CRETACEOUS PERIODS, THE TABLE IS CLEARED FOR THE RADIATIONS OF **DINOSAURS** AND **GIANT REPTILES**, SOME OF THE **COOLEST** ORGANISMS **EVER**!

Bloort, do I detect a distinct lack of scientific objectivity?

I'M SORRY, YOUR MAJESTY, BUT EVER SINCE I LEARNED ABOUT THESE CREATURES, MY IMAGINATION HAS BEEN THRILLED BY THEIR WONDROUS VARIETY AND SIZE. YOU SEE, THEY'RE...THEY'RE...WELL, THEY'RE REALLY, REALLY **BIG**!

Get a hold of yourself, Bloort! I have trouble believing that any nonsquinch organism could possible be that...

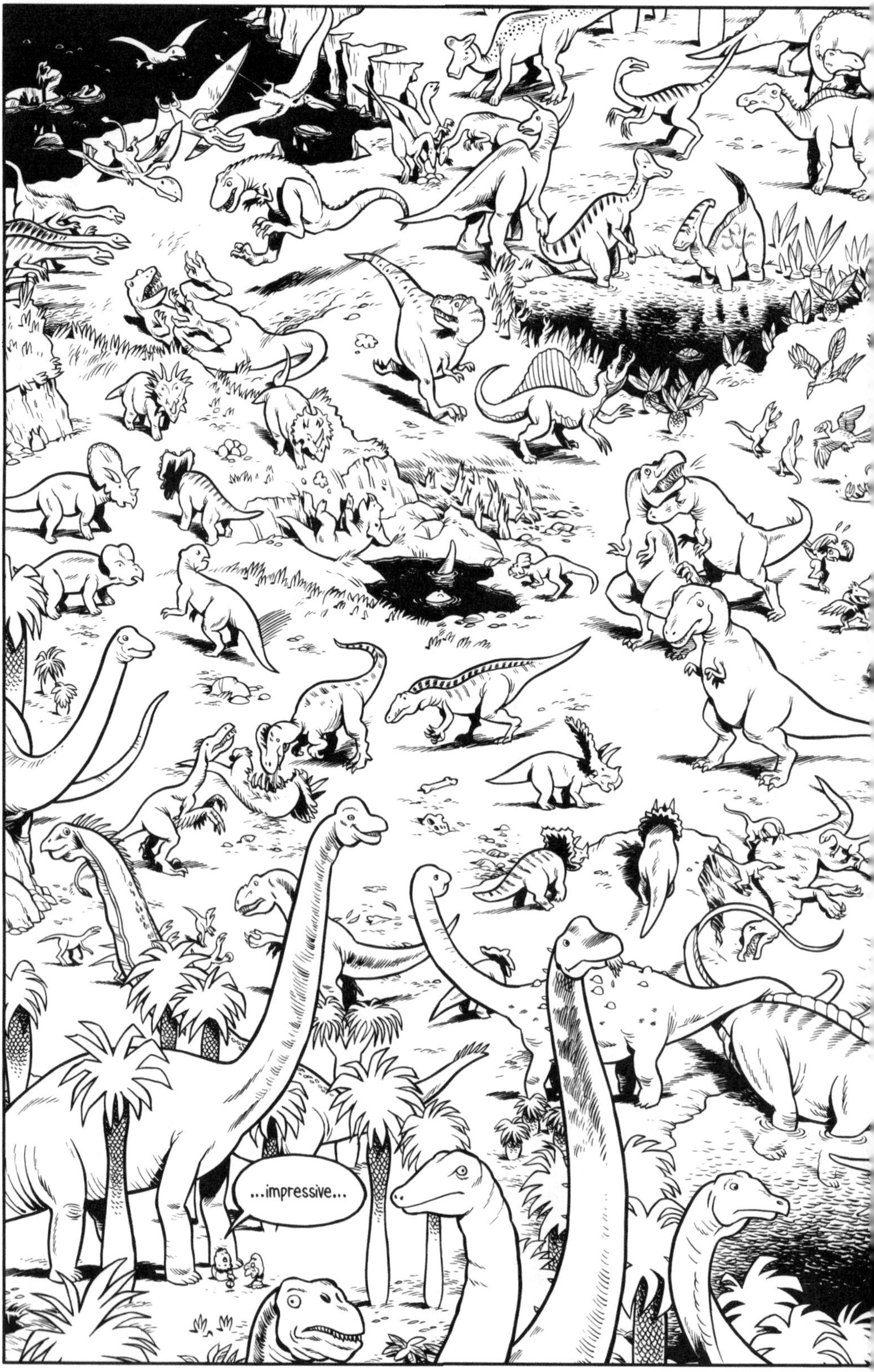
...impressive...

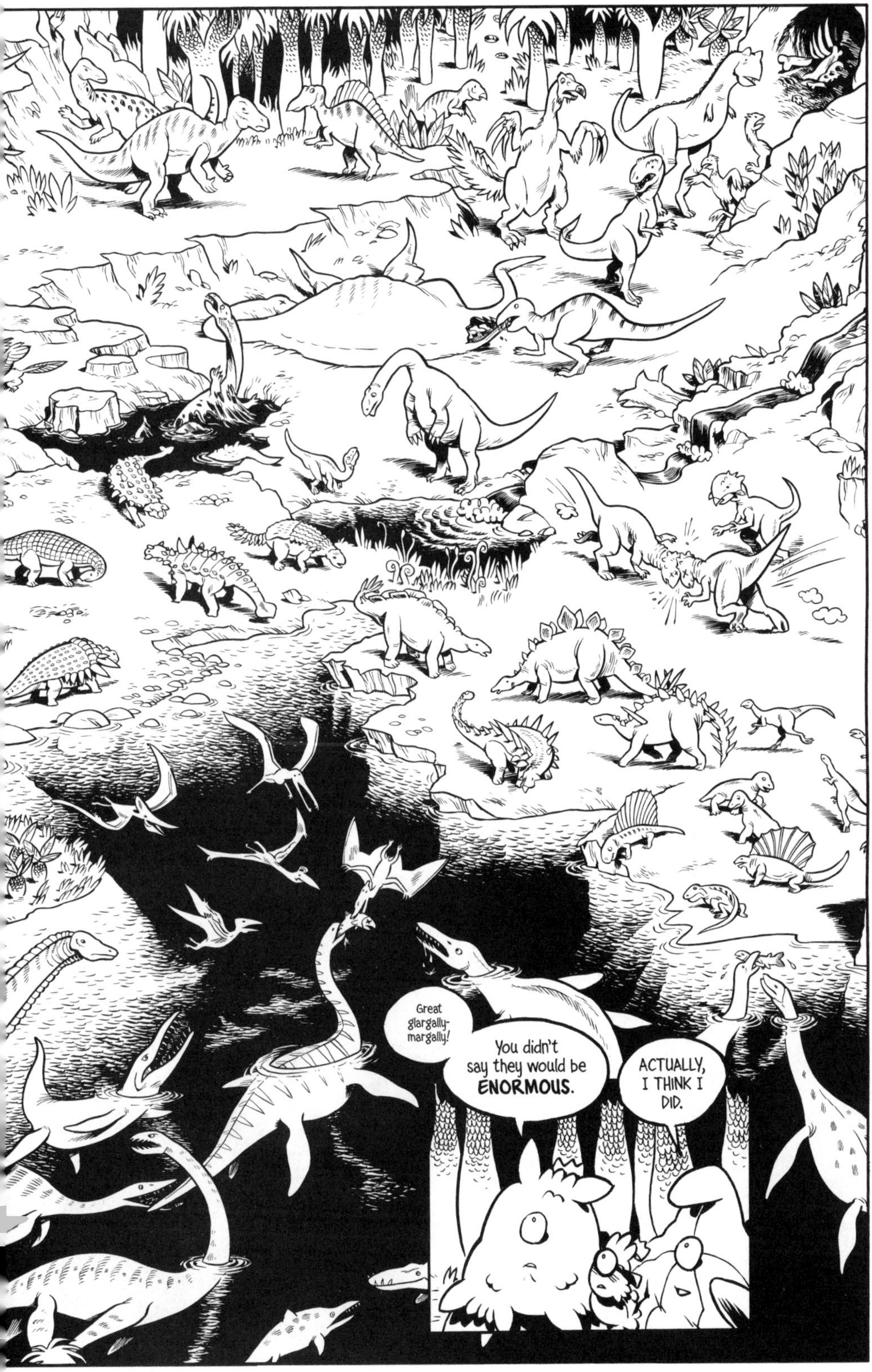
Great glargally-margally!
You didn't say they would be **ENORMOUS**.
ACTUALLY, I THINK I DID.

I can certainly see what you mean, Bloort, those are -- **EEP!**

What was **THAT**?

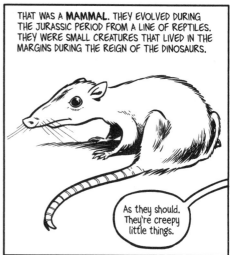

THAT WAS A **MAMMAL**. THEY EVOLVED DURING THE JURASSIC PERIOD FROM A LINE OF REPTILES. THEY WERE SMALL CREATURES THAT LIVED IN THE MARGINS DURING THE REIGN OF THE DINOSAURS.

As they should. They're creepy little things.

Bloort! This big mammal is being eaten by these smaller ones.

DON'T BE ALARMED, YOUR HIGHNESS. YOU'RE WITNESSING A DEFINING CHARACTERISTIC OF THIS GROUP OF ANIMALS.

Cannibalism?

No, **NURSING**. FEMALE MAMMALS HAVE **MAMMARY GLANDS** THAT PRODUCE A NUTRIENT-RICH FLUID CALLED MILK, WHICH THEIR YOUNG FEED UPON.

Eww.

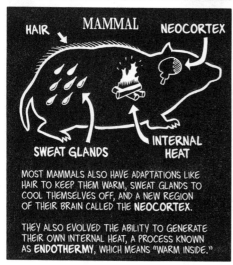

HAIR

MAMMAL

NEOCORTEX

SWEAT GLANDS

INTERNAL HEAT

MOST MAMMALS ALSO HAVE ADAPTATIONS LIKE HAIR TO KEEP THEM WARM, SWEAT GLANDS TO COOL THEMSELVES OFF, AND A NEW REGION OF THEIR BRAIN CALLED THE **NEOCORTEX**.

THEY ALSO EVOLVED THE ABILITY TO GENERATE THEIR OWN INTERNAL HEAT, A PROCESS KNOWN AS **ENDOTHERMY**, WHICH MEANS "WARM INSIDE."

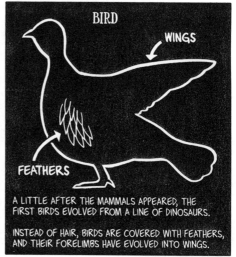

BIRD

WINGS

FEATHERS

A LITTLE AFTER THE MAMMALS APPEARED, THE FIRST BIRDS EVOLVED FROM A LINE OF DINOSAURS.

INSTEAD OF HAIR, BIRDS ARE COVERED WITH FEATHERS, AND THEIR FORELIMBS HAVE EVOLVED INTO WINGS.

IMAGINE A WORLD WHERE BIG BIRD WAS KING AND YOU WILL HAVE A PRETTY GOOD PICTURE OF WHAT THE ISLANDS OF NEW ZEALAND WERE LIKE OVER SEVEN CENTURIES AGO.

NO MORE MOA

MORE THAN 250 DIFFERENT BIRD SPECIES POPULATED THE ISLANDS, INCLUDING ENORMOUS ANIMALS LIKE GIANT **MOAS** AND **HAAST'S EAGLES.** THIS HAVEN FOR BIRDS EXISTED FOR MILLIONS OF YEARS, UNTIL IT WAS RAPIDLY DECIMATED BY HUMANS SEVEN CENTURIES AGO.

THE ISLANDS OF NEW ZEALAND ARE PRETTY ISOLATED AND, NOT SURPRISINGLY, ANIMALS THAT COULD FLY WERE MOST LIKELY TO COLONIZE THEM. AS A RESULT, THERE WEREN'T ANY TERRESTRIAL, OR LAND-DWELLING, MAMMALS LIKE RATS OR CATS ON NEW ZEALAND, AND BIRDS WERE THE TOP DOGS (SO TO SPEAK). BECAUSE THEY DIDN'T HAVE TO COMPETE WITH MAMMALS FOR RESOURCES, MANY SPECIES OF NEW ZEALAND BIRDS EVOLVED TO **ENORMOUS** PROPORTIONS.

THE BIGGEST OF THE NEW ZEALAND BIRDS WERE THE GIANT MOAS THAT STOOD THIRTEEN FEET HIGH AND WEIGHED IN AT 500 POUNDS. THE MOAS WERE FLIGHTLESS BIRDS THAT ROAMED THE FOREST OF NEW ZEALAND, EATING PLANTS. FOR OVER A MILLION YEARS, THE ONLY THREAT TO THEIR EXISTENCE WAS A GIANT PREDATOR, HAAST'S EAGLE, WHICH COULD GROW TO BE AS MUCH AS THREE FEET HIGH, FIVE FEET LONG, AND HAVE A WINGSPAN OF OVER TEN FEET WIDE.

AROUND A.D. 1300, A TRIBE OF POLYNESIANS CALLED THE MAORI DISCOVERED NEW ZEALAND AND ESTABLISHED A SETTLEMENT. THE MAORI HUNTED AND ATE THE MOAS AND DESTROYED MUCH OF THEIR FOREST HABITAT. IN ADDITION, HUMANS INADVERTENTLY BROUGHT STOWAWAYS LIKE **RATS** WITH THEM WHEN THEY ARRIVED. THESE INTRODUCED RATS FEASTED ON THE MOAS' EGGS, WHICH SAT CONVENIENTLY IN NESTS ON THE GROUND. THE MOAS COULD NOT REPRODUCE FAST ENOUGH IN THE FACE OF THIS HUMAN AND RAT ONSLAUGHT. WITHIN ONE HUNDRED YEARS OF HUMAN ARRIVAL, THE MOAS HAD BEEN DRIVEN TO EXTINCTION. BUT THE MOAS DID NOT GO ALONE. WITHOUT THEIR PRIMARY FOOD SOURCE, HAAST'S EAGLES SOON FOLLOWED THE MOAS INTO EXTINCTION.

TODAY, MANY SPECIES ARE AT RISK OF EXTINCTION AS A RESULT OF HUMAN HUNTING AND FISHING. WHAT'S MORE, THE LOSS OF ONE SPECIES CAN RESULT IN THE LOSS OF MANY MORE. SCIENTISTS AND CONSERVATIONISTS ARE LOOKING FOR BETTER WAYS TO MANAGE EARTH'S RESOURCES SO THAT WE CAN AVOID THE WIDESPREAD LOSS OF SPECIES VITAL FOR HUMAN SURVIVAL AND PERHAPS PREVENT OUR OWN EXTINCTION.

LIKE MAMMALS, BIRDS EVOLVED ENDOTHERMY. AS WE WILL SEE, THIS IS A VERY NICE TRAIT TO HAVE SHOULD THE EARTH EVER EXPERIENCE A CATASTROPHIC DROP IN GLOBAL TEMPERATURES.

uh... that sounds an awful lot like foreshadowing, Bloort.

AS THE NUMBER OF DINOSAURS BLOOMED IN THE LATE CRETACEOUS, SO TOO DID THE FIRST FLOWERING PLANTS. FLOWERING PLANTS, ALSO KNOWN AS **ANGIO-SPERMS**, RADIATED RAPIDLY AT THE END OF THE CRETACEOUS WITH THE HELP OF INSECTS LIKE BEES SPREADING THEIR POLLEN FROM FLOWER TO FLOWER.

HONEY, I... I'VE BEEN WITH ANOTHER FLOWER.

AGAIN?

CAN'T YOU SEE THIS IS **POLLEN** ME APART?

Wait a minute, Bloort, I've been thinking. At the beginning of this extinction business you said there had been **FIVE** mass extinctions. The trilobites survived the first two, but not the third in the Permian. That makes the Triassic the **FOURTH**.

CORRECT, YOUR MOST PERCEPTIVE GRAND HEAVINESS.

These magnificent creatures aren't going to make it, are they?

NO.

BUT, THEIR DESCENDANTS -- THE BIRDS -- WILL.

And did humans evolve from birds?

NO.

So, what **DID** they --

SCURRY

You're kidding me.

NOPE.

AT THE END OF THE CRETACEOUS, AN ASTEROID THE SIZE OF A MOUNTAIN STRUCK EARTH OFF THE COAST OF MEXICO.

THE IMPACT TRIGGERED DEVASTATING EARTHQUAKES, TIDAL WAVES, ACID RAIN, AND VOLCANIC ACTIVITY. THE HEAT FROM THE BLAST IGNITED WILDFIRES ACROSS THE GLOBE AND DEVASTATED FORESTS.

AFTER THE FIRES SUBSIDED, THE ASH AND SMOKE THEY PRODUCED BLOTTED OUT THE SUN AND LEFT THE WORLD COLD AND DARK.

THE RESULT WAS A **CALAMITY** FOR THE WORLD'S ECOSYSTEMS.

AN **ECOSYSTEM** IS A COMMUNITY THAT RESULTS FROM THE INTER-ACTION BETWEEN ORGANISMS AND THEIR PHYSICAL ENVIRONMENT.

AS PLANT AND ANIMAL SPECIES WENT EXTINCT, THOSE COMPLEX INTERACTIONS **COLLAPSED**, CAUSING EVEN MORE SPECIES TO GO EXTINCT. DINOSAURS, FLYING REPTILES, AND MARINE REPTILES WERE COMPLETELY WIPED OUT.

NEVERMORE.

YOU'VE BEEN **DYING** TO SAY THAT, HAVEN'T YOU?

CENOZOIC ERA

IN THE CHILLY CLIMATE AFTER THE ASTEROID STRIKE, THE WARM LITTLE MAMMALS AND BIRDS SCAMPERED AND FLEW FROM THE PERIPHERY TO RADIATE INTO THE NICHES LEFT VACANT BY THE DINOSAURS.

ON LAND, THEY EVOLVED INTO A VARIETY OF PREDATORS AND PLANT-EATERS, INCLUDING THE LARGEST MAMMALS AND BIRDS THAT EVER EXISTED.

SOME MAMMALS RETURNED TO THE SEA, EVOLVING STREAMLINED FORMS JUST AS THE ICHTHYOSAURS HAD DONE ALMOST 200 MILLION YEARS EARLIER.

WE LIKE TO EAT FISH, **TOO**, Y'KNOW.

STILL ANOTHER LINE OF MAMMALS, CALLED **PRIMATES**, WOULD TAKE TO THE TREES AND RAPIDLY DIVERSIFY.

ABOUT 30 MILLION YEARS AGO, A GROUP OF PRIMATES CALLED **APES** DESCENDED FROM THE TREES...

...ABOUT 200,000 YEARS AGO, A LINE OF HOMINIDS EVOLVED INTO **HUMANS**...

...AND FORTY YEARS AGO, HUMANS SENT MEMBERS OF THEIR SPECIES TO EARTH'S MOON.

That was a bit **RUSHED**, wasn't it, Bloort?

PERHAPS, MY MOST MOMENTOUS MONARCH, BUT WE'LL BE SPENDING MUCH MORE TIME WITH THE HUMANS LATER.

Later? But I have a meeting in a few...

So, how do **WE** survive extinction, Bloort?

We **ARE** facing a genetic crisis, aren't we?

I...UH.

Have you been listening in on my conversations again?

Gee, Dad, can I help it if your voice carries?

Look, it just seems to me that we could learn something from the fact that life on Earth has a four-and-a-half-billion-year history of surviving catastrophe.

Very perceptive, Son. Any ideas, Bloort?

THE PRINCE CONTINUES TO DAZZLE WITH HIS PROFOUND PERCEPTIVENESS, MY LORD. BUT I DON'T THINK MASS EXTINCTION IS THE MAJOR CONCERN HERE.

EARTH SCIENTISTS PROJECT THAT UP TO **99.9%** OF ALL SPECIES THAT HAVE EVER EXISTED ON EARTH ARE NOW EXTINCT.

WHILE MASS EXTINCTIONS LIKE THE ONES WE'VE JUST SEEN ARE DRAMATIC, THEY ACCOUNT FOR ONLY 4% OF ALL THE EXTINCT SPECIES IN THE HISTORY OF LIFE. THE REMAINING 96% OF ALL EXTINCTIONS ON EARTH ARE CALLED **BACKGROUND EXTINCTIONS.**

LOOK AT THOSE SPOTTED ONES. FEH. SO COMMON.

BUT WE'RE SPECIAL BECAUSE **WE** ARE THE LAST TWO OF OUR SPECIES.

BACKGROUND EXTINCTIONS ARE THE CONTINUOUS, LOW-LEVEL LOSS OF SPECIES THROUGH THE NORMAL COURSE OF EVOLUTION. AND WHILE MASS EXTINCTIONS TEND TO HAVE RAPID, GLOBAL EFFECTS ON A WIDE RANGE OF SPECIES, BACKGROUND EXTINCTIONS OCCUR OVER SMALLER, MORE LOCAL AREAS.

SPLORT!!

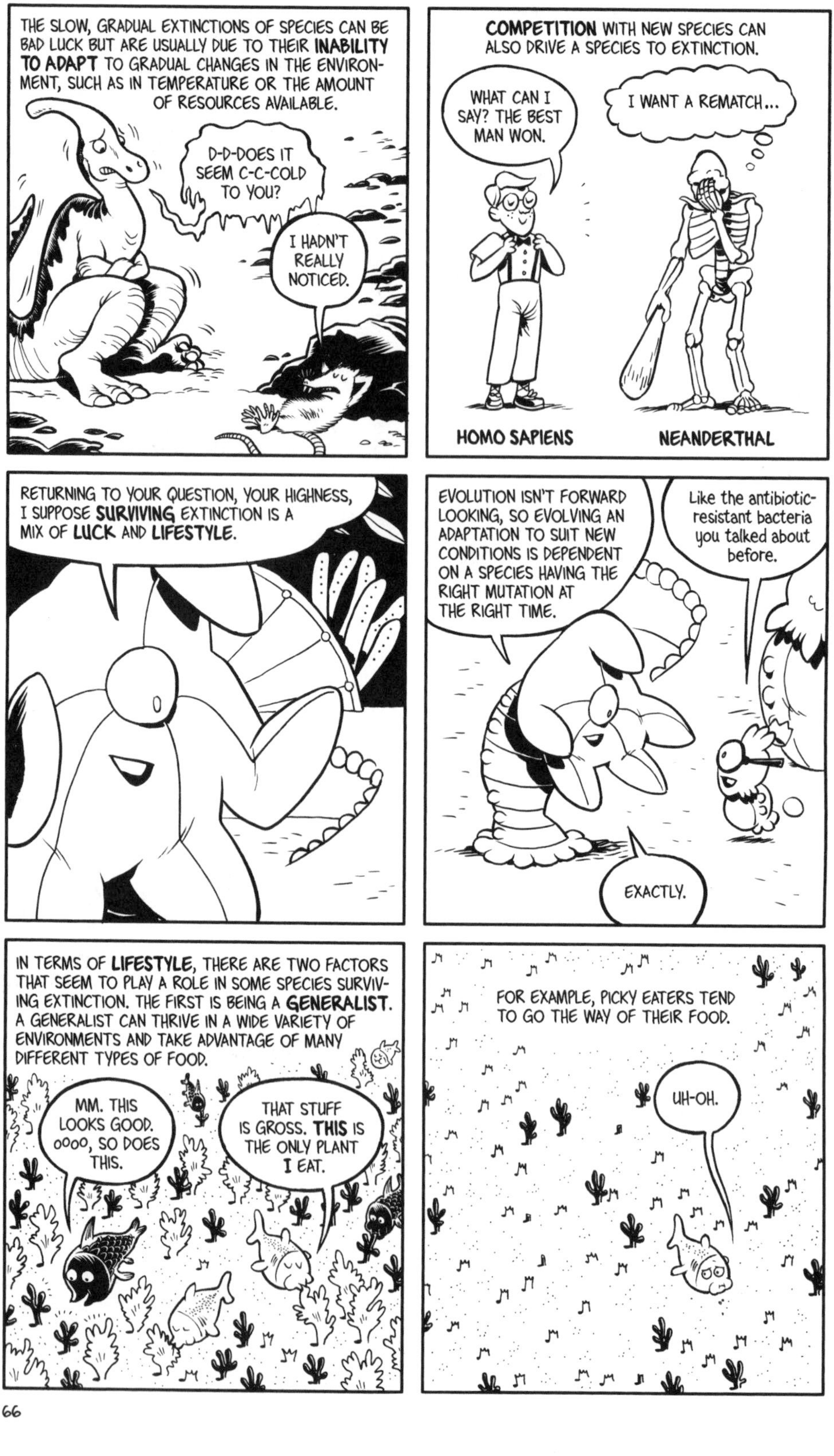
THE SLOW, GRADUAL EXTINCTIONS OF SPECIES CAN BE BAD LUCK BUT ARE USUALLY DUE TO THEIR **INABILITY TO ADAPT** TO GRADUAL CHANGES IN THE ENVIRONMENT, SUCH AS IN TEMPERATURE OR THE AMOUNT OF RESOURCES AVAILABLE.
D-D-DOES IT SEEM C-C-COLD TO YOU?
I HADN'T REALLY NOTICED.
COMPETITION WITH NEW SPECIES CAN ALSO DRIVE A SPECIES TO EXTINCTION.
WHAT CAN I SAY? THE BEST MAN WON.
I WANT A REMATCH...
HOMO SAPIENS
NEANDERTHAL
RETURNING TO YOUR QUESTION, YOUR HIGHNESS, I SUPPOSE **SURVIVING** EXTINCTION IS A MIX OF **LUCK** AND **LIFESTYLE**.
EVOLUTION ISN'T FORWARD LOOKING, SO EVOLVING AN ADAPTATION TO SUIT NEW CONDITIONS IS DEPENDENT ON A SPECIES HAVING THE RIGHT MUTATION AT THE RIGHT TIME.
Like the antibiotic-resistant bacteria you talked about before.
EXACTLY.
IN TERMS OF **LIFESTYLE**, THERE ARE TWO FACTORS THAT SEEM TO PLAY A ROLE IN SOME SPECIES SURVIVING EXTINCTION. THE FIRST IS BEING A **GENERALIST**. A GENERALIST CAN THRIVE IN A WIDE VARIETY OF ENVIRONMENTS AND TAKE ADVANTAGE OF MANY DIFFERENT TYPES OF FOOD.
MM. THIS LOOKS GOOD. OOOO, SO DOES THIS.
THAT STUFF IS GROSS. **THIS** IS THE ONLY PLANT **I** EAT.
FOR EXAMPLE, PICKY EATERS TEND TO GO THE WAY OF THEIR FOOD.
UH-OH.

A good thing to keep that in mind next time we have Glargalian Worm Noodles.

Bleh.

THE SECOND LIFESTYLE FACTOR MAY BE AN ORGANISM'S **GEOGRAPHIC DISTRIBUTION**.

THE EARTH SCIENTIST DAVID JABLONSKI HAS BEEN USING CLAMS TO STUDY EXTINCTION. THESE ARE GREAT ORGANISMS TO STUDY BECAUSE THEY'VE BEEN AROUND A LONG TIME, HAVE HARD SHELLS, AND LIVE BURIED IN THE MUD.

So when they die, they are already in a prime location to fossilize.

WAY TO CONNECT THE DOTS, SIRE! YOUR ASTUTE OBSERVATIONS CONTINUE TO PROVIDE ILLUMINATION AND INSPIRATION TO THE TOUR.

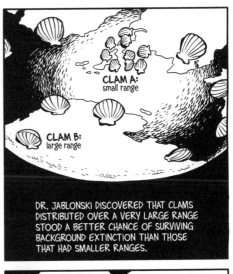

CLAM A: small range

CLAM B: large range

DR. JABLONSKI DISCOVERED THAT CLAMS DISTRIBUTED OVER A VERY LARGE RANGE STOOD A BETTER CHANCE OF SURVIVING BACKGROUND EXTINCTION THAN THOSE THAT HAD SMALLER RANGES.

I think I see. It would be like if some catastrophe destroyed the squinch colony on the Abyssal Plain. The squinch species would still persist because we have colonies at many different levels of depth in the ocean.

INDEED, YOUR HIGHNESS.

Although let's not share that example with the Ambassador from Abyssal Colony, okay?

What we really need now is to develop a better understanding of how species arise and adapt to new situations. If we know that, we might be able to...

But I guess you've gotta go to your meeting.

SIGH. Thanks, Bloort. I had a really great time.

THE PLEASURE WAS MINE, YOUR HIGHNESS.

Just a moment, Son, I need to make a call.

Chamberlain? I am engaged in a critical scientific investigation.

Cancel my meeting for today.

Thanks, Dad.

Yes, well, I am the **KING**, you know.

CHAPTER 4

Old Things into New

WE'VE JUST HAD THE OPPORTUNITY TO SEE A FOUR-AND-A-HALF-BILLION-YEAR PARADE OF NEW SPECIES.

THE QUESTION NOW IS **HOW** DO NEW SPECIES EVOLVE? TO ANSWER THAT, THIS PART OF THE TOUR WILL FOCUS ON THE CONCEPT OF SPECIATION.

SPECIATION IS THE PROCESS THROUGH WHICH ONE SPECIES EVOLVES INTO ONE OR TWO NEW, DISTINCT SPECIES. THIS HAS HAPPENED REPEATEDLY DURING THE HISTORY OF LIFE ON EARTH AND THERE ARE LOTS OF COOL EXAMPLES I CAN'T WAIT TO SHOW Y--

Bloort?

YES, MY MOST PRECOCIOUS PRINCE.

What's a **SPECIES**? I mean, how do we know when we have a new one?

I AM GRATEFUL THAT YOUR MOST PRODIGIOUS INTELLECT HAS STRUCK UPON A KEY ELEMENT THUS FAR MISSING IN THE PRESENTATION. AS YOU SUGGEST...

...BEFORE WE CAN DISCUSS SPECIATION, WE NEED TO DEFINE WHAT A SPECIES IS.

THE BIOLOGIST ERNST MAYR CREATED THE MOST COMMONLY USED DEFINITION. IT'S CALLED THE **BIOLOGICAL SPECIES CONCEPT**.

"Species are groups of actually or potentially interbreeding populations, which are reproductively isolated from other such groups."

MAYR

IN OTHER WORDS, A SPECIES IS A GROUP OF ORGANISMS THAT SHARE COMMON CHARACTERISTICS AND WHO MATE WITH EACH OTHER IN NATURE, BUT DO NOT MATE WITH OTHER SIMILAR GROUPS OF ORGANISMS.

DIDN'T I JUST SAY THAT?

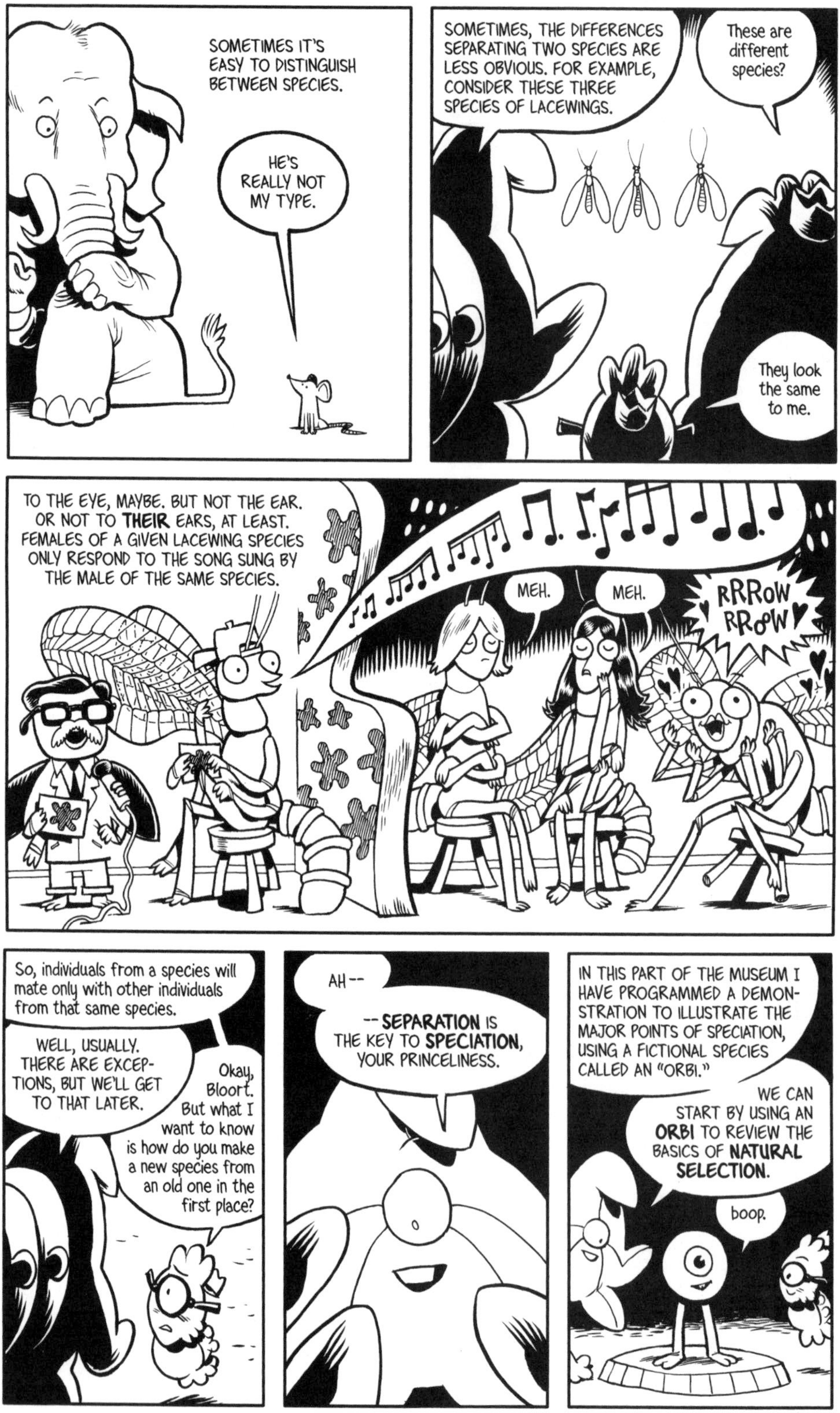
SOMETIMES IT'S EASY TO DISTINGUISH BETWEEN SPECIES.
HE'S REALLY NOT MY TYPE.
SOMETIMES, THE DIFFERENCES SEPARATING TWO SPECIES ARE LESS OBVIOUS. FOR EXAMPLE, CONSIDER THESE THREE SPECIES OF LACEWINGS.
These are different species?
They look the same to me.
TO THE EYE, MAYBE. BUT NOT THE EAR. OR NOT TO THEIR EARS, AT LEAST. FEMALES OF A GIVEN LACEWING SPECIES ONLY RESPOND TO THE SONG SUNG BY THE MALE OF THE SAME SPECIES.
MEH.
MEH.
RRROW RROW
So, individuals from a species will mate only with other individuals from that same species.
WELL, USUALLY. THERE ARE EXCEPTIONS, BUT WE'LL GET TO THAT LATER.
Okay, Bloort. But what I want to know is how do you make a new species from an old one in the first place?
AH--
--SEPARATION IS THE KEY TO SPECIATION, YOUR PRINCELINESS.
IN THIS PART OF THE MUSEUM I HAVE PROGRAMMED A DEMONSTRATION TO ILLUSTRATE THE MAJOR POINTS OF SPECIATION, USING A FICTIONAL SPECIES CALLED AN "ORBI."
WE CAN START BY USING AN ORBI TO REVIEW THE BASICS OF NATURAL SELECTION.
boop.

CAN YOU DESCRIBE ITS **PHENOTYPE** FOR ME, YOUR HIGHNESS?

It has one eye, two legs, a round body, smooth skin, and two little teeth in front. Oh, and it keeps saying "boop."

EXCELLENT! THAT IS EXACTLY THE ORBI'S **PHENOTYPE** -- ITS PHYSICAL AND BEHAVIORAL CHARACTERISTICS.

Wait...I know this stuff, too. Those physical characteristics are determined by the creature's **GENOTYPE** -- the unique set of genes the organism has.

Nice one, Dad.

I'm all over this genetics stuff.

HIGH FIVE!

YES YOU ARE, MOST PERCEPTIVE OMNISCIENCE.

AS IN ANY POPULATION, WE SEE VARIATION IN ORBI **PHENOTYPES**. SOME ARE SLIGHTLY TALLER THAN OTHERS, FOR EXAMPLE.

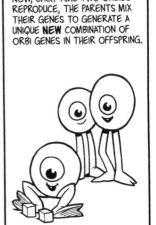

NOW, EACH TIME TWO ORBIES REPRODUCE, THE PARENTS MIX THEIR GENES TO GENERATE A UNIQUE **NEW** COMBINATION OF ORBI GENES IN THEIR OFFSPRING.

BUT THEY CAN ONLY MIX THE GENES THAT ARE **AVAILABLE**. CONSEQUENTLY, SINCE NONE OF THE ORBIS HAVE WINGS, THERE ARE NO WING GENES TO MIX AND SHARE.

They are limited to the genetic information present in the orbi population.

EXACTLY. WE CALL THE SUM OF A POPULATION'S GENETIC INFORMATION ITS **GENE POOL**.

Which means if you go fishing in the orbi gene pool, you might catch genes for leg length, but you won't be reeling in any wing genes.

AN ASTUTE AND ENGAGING WAY TO PUT IT, YOUR HIGHNESS. IT TRULY IS AN EXHILARATING EXPERIENCE TO BE IN THE PRESENCE OF SUCH LUMINOUS INTELLECTS.

BOW!

You know, Bloort, if you continue with the elaborate compliments we'll never get through this tour.

Never complain about the groveling of underlings, son.

I WILL ATTEMPT TO STRIKE A MORE BALANCED APPROACH TO MY OBSEQUIOUSNESS, YOUR HIGHNESS.

OUR ORBIES MIGHT STAY LIKE THIS FOR A VERY LONG TIME IF THE GENE POOL DOESN'T CHANGE. THIS, OF COURSE, IS THE VERY GENETIC CRISIS WE SQUINCH NOW FACE AS A SPECIES.

BUT LET'S IMAGINE WHAT MIGHT HAPPEN IF A MUTATION AROSE THAT GAVE SOME ORBIES LITTLE TUFTS OF HAIR ALL OVER THEIR BODIES. AT FIRST IT MAY SEEM THAT ALL THOSE TUFTS ARE ONLY GOOD FOR MAKING OTHER ORBIES LAUGH.

HA. Boop.

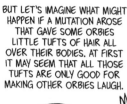

IT SOON BECOMES CLEAR, HOWEVER, THAT THE ORBIES WITH HAIR TUFTS ARE WELL CAMOUFLAGED FROM THE HORNOZZLE ORBI-EATER.

PREDATION IS A STRONG FORCE FOR NATURAL SELECTION. BECAUSE OF PRESSURE FROM THE HORNOZZLE ORBI-EATER, THE ORBIES WITH THE TUFT GENE HAVE AN INCREASED LIKELIHOOD OF SURVIVAL AND THEREFORE OF LEAVING MORE OFFSPRING THAN THEIR BALD COUNTERPARTS.

NOW THE POPULATION CONTAINS BOTH TUFTED AND BALD PHENOTYPES.

Are they different species yet?

NOT YET, SINCE BOTH PHENOTYPES CAN STILL FREELY MATE AND THE GENES FOR BOTH TRAITS ARE MAINTAINED IN THE POPULATION. REMEMBER THAT I SAID THE KEY TO SPECIATION WAS **SEPARATION**.

Yes, and the suspense is killing me. How does separation happen?

THERE ARE A NUMBER OF WAYS, BUT PERHAPS THE EASIEST TO PICTURE IS GEOGRAPHIC ISOLATION. IN **GEOGRAPHIC ISOLATION**, TWO POPULATIONS ARE SEPARATED BY A PHYSICAL BARRIER, SUCH AS AN OCEAN, MOUNTAIN, OR CANYON.

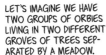

LET'S IMAGINE WE HAVE TWO GROUPS OF ORBIES LIVING IN TWO DIFFERENT GROVES OF TREES SEPARATED BY A MEADOW.

THE ORBIES FROM BOTH GROUPS ARE FREE TO MATE WITH EACH OTHER AND MIX THEIR GENES. BECAUSE OF THIS, THERE IS GENE FLOW BETWEEN THE TWO SMALL POPULATIONS.

GENE FLOW IS THE PROCESS BY WHICH GENES IN ONE GENE POOL CAN MOVE TO ANOTHER THROUGH REPRODUCTION.

GENE FLOW

NOW IMAGINE THAT A MOUNTAIN RANGE SUDDENLY SPRINGS UP IN THE MEADOW AND ISOLATES THE TWO ORBI POPULATIONS FROM EACH OTHER.

SINCE THE TWO POPULATIONS ARE NOW COMPLETELY SEPARATED, THE ORBIES CANNOT BREED AND MIX THEIR GENES. UNDER THESE CONDITIONS WE SAY THAT THERE IS **NO** GENE FLOW BETWEEN THEIR GENE POOLS.

IF WE KEEP THEM SEPARATED FOR A LONG TIME, EACH POPULATION WILL FOLLOW A SEPARATE EVOLUTIONARY PATH AS THEY EVOLVE ADAPTATIONS TO THEIR UNIQUE ENVIRONMENTAL CONDITIONS.

THE **WESTERN** SIDE OF THE MOUNTAIN FACES THE SEA, AND THE CLIMATE IS HOT AND HUMID. UNDER THESE CONDITIONS, EVOLUTION MIGHT FAVOR THE SELECTION OF BALD ORBIES THAT HAVE LONG LEGS TO ELEVATE THEM OFF THE HOT SAND.

ON THE **EASTERN** SIDE OF THE ORBI MOUNTAINS, THE AVERAGE TEMPERATURE BECOMES VERY COLD AND IT SNOWS A LOT. UNDER THESE CONDITIONS, EVOLUTION MIGHT FAVOR EASTERN ORBIES WITH TUFTS ALL OVER THEIR BODY AND SHORTER LEGS TO KEEP WARMER IN THIS CHILLY CLIMATE.

WESTERN ORBIES DINE ON FLYING INSECTS AND USE THEIR LONG LEGS TO LEAP INTO THE AIR AND CATCH PREY ON THE WING.

Boop.

EASTERN ORBIES EAT SMALL INSECTS IN THE SOIL, SO INDIVIDUALS WITH BIG, DIGGING TEETH HAVE A SELECTIVE ADVANTAGE.

Boop. **BRAIN FREEZE.** Boop. Boop!

EVENTUALLY, THE ORBIES IN EACH OF THESE POPULATIONS MIGHT START USING THE SURVIVAL VALUE OF THESE TRAITS TO DECIDE WHO IS BEST TO MATE WITH. FOLLOW EACH ISOLATED GROUP LONG ENOUGH, AND YOU MIGHT SEE SOMETHING LIKE THE FOLLOWING...

Boop. Boop.

ORBIES IN THE WEST EVOLVE ELABORATE LEG DANCES TO SHOW-OFF THEIR ELONGATED APPENDAGES. PERHAPS SELECTION FAVORS THE EVOLUTION OF COLORFUL LEG BANDS THAT ACCENTUATE THE DISPLAY. LET'S CALL THESE **BANDED ORBIES.**

Boop.

EASTERN ORBIES, ON THE OTHER HAND, EVOLVE AN ENERGETIC SHAKING DISPLAY THAT SHOWCASES THEIR THICK COAT AND BIG TEETH TO POTENTIAL MATES. WE'LL CALL THESE **WOOLLY ORBIES.**

Boop.

HERE IS WHY THE SEPARATION IS IMPORTANT: GENES FOR ADAPTATIONS THAT EVOLVE IN ONE POPULATION CANNOT GET INTO THE GENE POOL OF THE OTHER POPULATION BECAUSE THE POPULATIONS AREN'T BREEDING AND MIXING THEIR GENES.

AFTER A MILLION YEARS OR SO, THE TWO POPULATIONS BECOME MORE AND MORE DIFFERENT. SINCE THERE IS NO GENE FLOW BETWEEN THE POPULATIONS, EACH HAS ACCUMULATED ITS OWN UNIQUE SET OF GENES AND ADAPTATIONS.

Are they different species now?

THE ONLY WAY TO KNOW THAT IS TO PUT THEM TOGETHER AND SEE IF THEY SUCCESSFULLY BREED.

? Boop? Boop? ?

They look like different species to me.

I DON'T THINK THEY ARE GOING TO MATE, THAT'S FOR SURE.

Boop.

Boop.

That is all well and good for a made-up population of holographic organisms, Bloort. But does this actually happen on Earth? As a **PRACTICAL** sovereign, I prefer to deal in **FACTUAL** information.

WOULD SOME SPECIFIC TERMS AND REAL-WORLD EXAMPLES HELP, YOUR MAJESTY?

Immensely.

THEN LET US BEGIN WITH SEXUAL REPRODUCTION.

AGAIN?

Good grief, Bloort, didn't you discuss the topic of Earth reproduction **ENOUGH** in your last report? You seem a bit obsessed with the topic. Frankly, I find the...process...of organisms mixing their genes **UNSETTLING**.

75

I APOLOGIZE, YOUR MAJESTY, BUT IT IS VITALLY IMPORTANT TO THIS DISCUSSION.

MULTICELLED EARTH ORGANISMS USUALLY COME IN TWO MAJOR FORMS: A **FEMALE** AND A **MALE**.

RHINOCEROS BEETLE

MALE   FEMALE

WHEN THEY SEEK TO PRODUCE OFFSPRING, THE FEMALE PRODUCES AN EGG THAT CONTAINS HALF OF HER GENES AND THE MALE PRODUCES A SPERM, WHICH CONTAINS HALF OF HIS GENES.

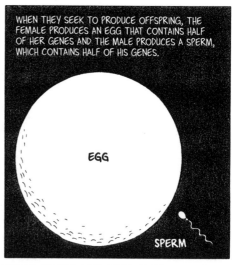

EGG

SPERM

WHEN THE SPERM AND EGG FUSE, THEY PRODUCE A **ZYGOTE**. THIS IS A SINGLE CELL THAT CONTAINS THE COMPLETE SET OF GENES FROM BOTH AND CAN GROW INTO AN ADULT ORGANISM.

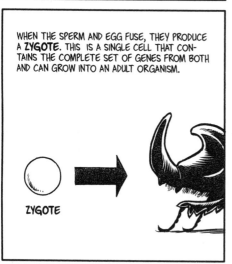

ZYGOTE

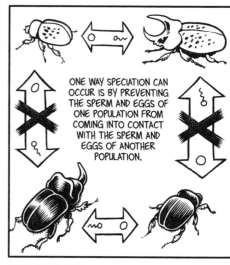

ONE WAY SPECIATION CAN OCCUR IS BY PREVENTING THE SPERM AND EGGS OF ONE POPULATION FROM COMING INTO CONTACT WITH THE SPERM AND EGGS OF ANOTHER POPULATION.

Of course. If you isolate two populations from each other they can't exchange sperm and eggs. That would effectively block gene flow.

WELL SAID, YOUR HIGHNESS. AS IT TURNS OUT, POPULATIONS CAN BECOME ISOLATED IN THREE BASIC WAYS: GEOGRAPHICALLY, ECOLOGICALLY, OR BEHAVIORALLY.

ALFRED RUSSEL WALLACE, DARWIN'S CONTEMPORARY AND CO-AUTHOR OF THE THEORY OF NATURAL SELECTION, NOTICED THAT THE AMAZON RIVER WAS AN EFFECTIVE GEOGRAPHIC BARRIER FOR **MARMOSETS**.

WALLACE

MARMOSETS ARE **PRIMATES**—A GROUP OF ORGANISMS THAT INCLUDE MONKEYS, GORILLAS, CHIMPANZEES, AND HUMANS. AS A RULE, PRIMATES DON'T LIKE TO SWIM.

THERE ARE EXCEPTIONS TO THAT RULE.

SINCE THE AMAZON CAN BE UP TO SEVEN MILES WIDE, AND MARMOSETS ON EITHER SIDE ARE RELUCTANT TO SWIM, THEY ARE UNLIKELY TO EVER ENCOUNTER EACH OTHER: **GEOGRAPHICAL ISOLATION.**

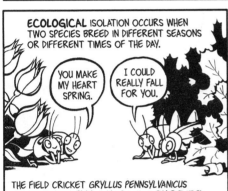

ECOLOGICAL ISOLATION OCCURS WHEN TWO SPECIES BREED IN DIFFERENT SEASONS OR DIFFERENT TIMES OF THE DAY.

YOU MAKE MY HEART SPRING.

I COULD REALLY FALL FOR YOU.

THE FIELD CRICKET *GRYLLUS PENNSYLVANICUS* REPRODUCES IN **FALL**, WHILE A CLOSELY RELATED SPECIES, *GRYLLUS VELETIS*, MATES IN THE **SPRING**.

BY MATING AT DIFFERENT TIMES OF THE YEAR THEY RARELY IF EVER ENCOUNTER EACH OTHER AND THERE IS NO GENE FLOW BETWEEN THE POPULATIONS.

BEHAVIORAL ISOLATION OCCURS WHEN MALES AND FEMALES OF A SPECIES WILL MATE ONLY WITH INDIVIDUALS WHO LOOK OR ACT A CERTAIN WAY.

Well, that seems kind of shallow. What if they have a really nice personality?

HMM, YES, I CAN SEE HOW THAT MIGHT BE CONFUSING FOR YOU.

Bloort?

I'M JUST WONDERING IF THIS WOULD BE A GOOD TIME TO BRING UP "SEXUAL SELECTION."

MORE details about Earthly reproduction?

THIS PART IS **CRUCIAL**, SIRE.

sigh

Very well.

REMEMBER THE CONCEPT OF FEMALES AND MALES, MY MOST SAGACIOUS PRINCE?

Yes. Having two sexes sounds really **WEIRD**.

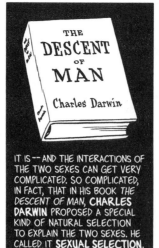

THE **DESCENT** OF **MAN**

Charles Darwin

IT IS -- AND THE INTERACTIONS OF THE TWO SEXES CAN GET VERY COMPLICATED. SO COMPLICATED, IN FACT, THAT IN HIS BOOK *THE DESCENT OF MAN*, **CHARLES DARWIN** PROPOSED A SPECIAL KIND OF NATURAL SELECTION TO EXPLAIN THE TWO SEXES. HE CALLED IT **SEXUAL SELECTION**.

IT IS THE PROCESS WHEREBY NATURE SELECTS FOR PHYSICAL OR BEHAVIORAL TRAITS THAT GIVE AN INDIVIDUAL AN **ADVANTAGE** IN **REPRODUCING**. MOSTLY, THIS REFERS TO TRAITS THAT HAVE EVOLVED IN MALES BECAUSE THE TRAITS ARE PREFERRED BY FEMALES FOR SOME REASON.

AS A RESULT, MALES AND FEMALES OF THE SAME SPECIES CAN OFTEN LOOK VERY DIFFERENT. TAKE THE **INDONESIAN BABIRUSA**, FOR EXAMPLE. THIS IS AN ANIMAL RELATED TO THE EARTHLY PIG AND, TO A LESSER EXTENT, THE HIPPOPOTAMUS.

IN BOTH MALES AND FEMALES, THE LOWER CANINE TEETH GROW INTO TUSKS.

YOU BOTH NEED TO START FLOSSING REGULARLY. AND, IF POSSIBLE, PLEASE, STOP FIGHTING WITH YOUR TEETH.

OINK.   OINK.

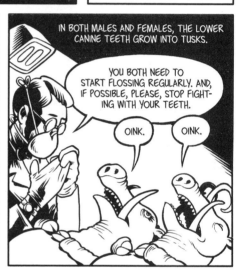

MALE BABIRUSA HAVE AN ADDITIONAL SET OF "ANTLERS" THAT FEMALES LACK. BUT IT TURNS OUT THAT THESE ANTLERS ARE ACTUALLY THE MALES' **UPPER CANINES** THAT GROW OUT OF THE TOP OF THE SKULL.

I'M AFRAID THIS MIGHT REQUIRE BRACES.

OINK.

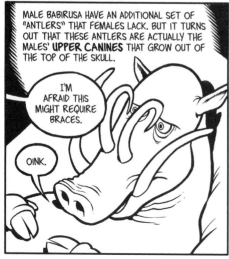

BLOOP!

FRANKLY, IT DOESN'T MAKE SENSE IN TERMS OF NATURAL SELECTION ALONE.

WHAT THE -- ? YOU'RE NOT IN THIS PART OF THE SIMULATION, MR. DARWIN.

I KNOW, BUT I WAS GETTING COOPED UP IN THAT COMPUTER THINGAMAJIG OF YOURS AND I NEEDED TO STRETCH MY LEGS.

SO, YOU REWROTE YOUR OWN PROGRAMMING?

IT WASN'T THAT HARD, SON. I DID SPEND A LIFETIME STUDYING THE PROCESS OF MODIFICATION, AFTER ALL.

NOW, WHERE WERE WE?

OH, YES, **THE MALE BABIRUSA'S ANTLERS.**

NATURAL SELECTION WOULD PREDICT THAT AN ANIMAL'S PHYSICAL TRAITS HAVE EVOLVED TO PROVIDE IT WITH SOME KIND OF **SURVIVAL ADVANTAGE**, RIGHT?

Yes, Sir.

SO, IF WE LOOK AT THE MALE BABIRUSA THROUGH THE LENS OF NATURAL SELECTION, WE SHOULD ASSUME THAT ALL OF HIS TRAITS, INCLUDING HIS TEETH-ANTLER-THINGEES, HELP HIM TO SURVIVE.

MALE BABIRUSA

Makes sense.

SO WHY DOESN'T THE FEMALE BABIRUSA HAVE TEETH GROWING OUT OF THE TOP OF **HER** HEAD? HER SURVIVAL IS VITAL FOR THE SURVIVAL OF THE SPECIES, TOO. AM I RIGHT?

FEMALE BABIRUSA

You **ARE** right. It isn't fair.

AH, BUT THIS ISN'T ABOUT **FAIR**, THIS IS ABOUT PASS-ING YOUR GENES ONTO THE NEXT GENERATION.

PERHAPS **I** SHOULD TAKE IT FROM HERE, SIR.

YES, EXCELLENT SUGGESTION, BLOORT. YOU'LL BE BETTER AT WORKING ALL THAT BUSINESS ABOUT **GENES** INTO THE EXPLANATION.

IN MANY CASES, DIFFERENCES BETWEEN MALES AND FEMALES EXIST BECAUSE THEY DON'T ALWAYS DO AN EQUAL AMOUNT OF WORK WHEN IT COMES TO MAKING BABIES. WE DON'T HAVE TO GO ANY FURTHER THAN THE EGG AND SPERM TO SEE THIS ILLUSTRATED.

THE FEMALE BABIRUSA INVESTS A LOT OF ENERGY INTO MAKING A BIG NUTRIENT-RICH EGG FULL OF ALMOST EVERYTHING THE ZYGOTE WILL NEED TO START DEVELOPING.

OKAY, LET'S SEE NOW. I'VE PACKED THE MITOCHONDRIA, LOTS OF ENERGETIC MOLECULES FOR WHEN WE GET HUNGRY...

BY CONTRAST, THE MALE BABIRUSA INVESTS MUCH LESS ENERGY TO MAKE HIS TINY SPERM CELLS, WHICH ONLY HAVE ENOUGH ROOM FOR THE MALE'S DNA AND A PROPELLER CALLED A **FLAGELLUM**.

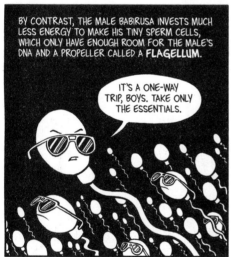

IT'S A ONE-WAY TRIP, BOYS. TAKE ONLY THE ESSENTIALS.

AFTER THEY MATE, THE FEMALE BABIRUSA WILL BE PREGNANT FOR ALMOST SIX MONTHS. IN THAT TIME SHE WILL EXPEND SIGNIFICANT AMOUNTS OF ENERGY GETTING FOOD FOR HER DEVELOPING PIGLETS, WHICH SHE WILL NURSE FOR AN ADDITIONAL SIX TO EIGHT MONTHS AFTER THEY'RE BORN.

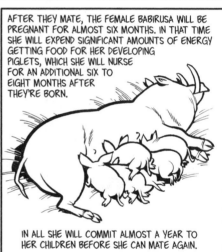

IN ALL SHE WILL COMMIT ALMOST A YEAR TO HER CHILDREN BEFORE SHE CAN MATE AGAIN.

Doesn't the male help?

NOT REALLY. ONCE HE HAS MATED, HE GOES LOOKING FOR ANOTHER MATE.

SINCE THE FEMALE IS GOING TO PUT SO MUCH **MORE** TIME AND ENERGY INTO NURTURING HER YOUNG, SHE WANTS TO MAKE SURE SHE MATES WITH A MALE WHO CAN GIVE HER BABIES THE BEST GENES FOR SURVIVAL.

IN THE CASE OF THE FEMALE BABIRUSA, SHE WANTS SOMEONE **TOUGH**.

I DON'T SEE WHAT THIS HAS TO DO WITH TEETH GROWING OUT OF THE TOP OF YOUR HEAD.

SMACK!

THE MALES STAND ON THEIR BACK LEGS AND BOX EACH OTHER USING THEIR LOWER TUSKS TO STAB THEIR RIVAL. THEIR ANTLER-TEETH ACT AS A SHIELD TO LOCK UP THEIR OPPONENT'S TUSKS.

NO RULES. NO LIMITS. SHAKE HOOVES AND COME OUT GOUGING.

THE ONE WHO WINS GETS TO MATE WITH THE FEMALE.

SINCE ONLY MALES WITH THE BIGGEST ANTLER-TEETH GET TO MATE, BIG ANTLER-TEETH GENES WILL BECOME MORE PROMINENT EACH GENERATION.

I assume this phenomenon is not restricted to **TEETH**.

INDEED, YOU ARE MAGNIFICENTLY INTUITIVE, SIRE. VIRTUALLY **ANY** TRAIT OR BEHAVIOR CAN BE MODIFIED BY SEXUAL SELECTION.

WHEN MALES COMPETE DIRECTLY WITH EACH OTHER FOR FEMALES, THEY CAN EVOLVE IMPRESSIVE ADAPTATIONS. HORNS, FOR EXAMPLE, HAVE EVOLVED NUMEROUS TIMES IN UNRELATED SPECIES AS A MEANS TO COMPETE FOR FEMALES.

BUT NOT ALL SEXUALLY SELECTED TRAITS EVOLVE FOR FIGHTING OTHER MALES. SOME EVOLVE BECAUSE THE MALE MUST CONVINCE THE FEMALE THAT HE IS THE BEST CHOICE FOR A MATE. CONSIDER THE BEHAVIOR DISPLAYS OF MALE **BOWERBIRDS**.

THERE ARE A TOTAL OF FIFTEEN DIFFERENT BOWERBIRD SPECIES IN AUSTRALIA AND ALL BUT TWO BUILD ELABORATE STRUCTURES, CALLED **BOWERS**, THAT THEY USE TO COURT FEMALES.

THE SATIN BOWERBIRD BUILDS AN AVENUE BOWER. AVENUE BOWERS CONSIST OF TWO PARALLEL ROWS OF TWIGS WITH A LITTLE AVENUE RUNNING DOWN THE MIDDLE.

THIS AVENUE IS THE PATHWAY TO MY HEART.

THE MALE DECORATES HIS BOWER WITH NEATLY ARRANGED STICKS AND COLORFUL ITEMS LIKE FLOWERS.

HMM, SHOULD THE BLUE RUBBER BAND BE ON THE LEFT OR THE RIGHT?

FEMALES LOOKING FOR A MATE WILL VISIT SEVERAL BOWERS TO INSPECT THE QUALITY AND TIDINESS OF THE DECORATIONS. DURING HER VISIT, THE MALE WILL TRY TO IMPRESS THE FEMALE WITH HIS SINGING AND DANCING.

DOO DOO DAH DOO, DO-DO, DO-DO, CAN'T TOUCH THIS!

I DON'T THINK I WANT TO.

DESPITE HIS BEST SONG AND DANCE, THE FEMALE RARELY STAYS TO SEE THE WHOLE SHOW DURING HER FIRST ROUND OF INSPECTIONS. SHE JUST WANTS TO GET A QUICK IMPRESSION OF WHAT THE MALES HAVE TO OFFER. WHEN SHE HAS SEEN ENOUGH, SHE GOES OFF TO BUILD A NEST.

A FEMALE SATIN BOWERBIRD TAKES ABOUT A WEEK TO BUILD HER NEST. WHEN SHE IS DONE, SHE RETURNS TO A FEW OF THE BOWERS SHE SAW THE WEEK BEFORE AND STAYS FOR THE WHOLE SHOW.

OH, GOOD, THERE'S A SEAT RIGHT IN FRONT.

IF SHE LIKES WHAT SHE SEES DURING THE STAGE SHOW, SHE WILL ACCOMPANY THE MALE INTO HIS BOWER, WHERE HE WILL REGALE HER WITH HIS IMPRESSIONS OF OTHER BIRD'S CALLS.

POLLY WANTS A CRACKER!

POLLY WANTS A CRACKER!

VERY AMUSING. VERY AMUSING INDEED.

IF SHE IS SUFFICIENTLY IMPRESSED WITH HIS ROUTINE, SHE'LL MATE WITH HIM. THEN SHE RETURNS TO HER NEST TO LAY HER EGGS AND RAISE HER YOUNG. SHE WON'T SEE THE MALE AGAIN.

I'M OKAY WITH BEING ALONE. THAT PARROT SCHTICK WOULD HAVE GOTTEN OLD REALLY FAST.

MEANWHILE, THE MALE STAYS IN HIS BOWER FOR THE ENTIRE TWO MONTHS OF THE BREEDING SEASON, TRYING TO WOW OTHER FEMALES WITH HIS SINGING, DANCING, AND IMPERSONATIONS.

THE SHOW MUST GO ON.

Wow, that's really an impressive set of behaviors, Bloort, but...um...well... **WHO CARES?**

BEG YOUR PARDON?

I mean, why should she care if the male can sing and dance and build a pretty nest? How does that help her pick the best mate?

CHOOSING THE MALE WITH THE BEST SONG AND DANCE HELPS THE FEMALE BECAUSE EACH OF THOSE BEHAVIORS ADVERTISE SOMETHING IMPORTANT ABOUT THE MALE.

BUILDING A GOOD BOWER IS PRETTY DEMANDING MENTAL ACTIVITY. DATA SUGGESTS THAT HEALTHIER MALES HAVE BIGGER BRAINS AND THEREFORE BUILD BETTER BOWERS. SO THE QUALITY OF THE BOWER MAY INDICATE THE MALE HAS **GOOD GENES** FOR GROWTH AND DEVELOPMENT.

LOOK AT THE CRANIUM ON **THAT** ONE.

BE STILL, MY BEATING HEART.

THE QUALITY OF THE BOWER AND THE NUMBER OF DECORATIONS ALSO TELL THE FEMALE HOW MANY **PARASITES** THE MALE HAS. MALES WITH MORE PARASITES TEND TO BUILD MESSIER BOWERS.

SO? WHATTAYA THINK OF MY PLACE?

WHEN HE DANCES, HE SHOWS OFF THE QUALITY OF HIS FEATHERS. FOR EXAMPLE, MALES WITH BRIGHT, COLORFUL FEATHERS HAVE FEWER BLOOD PARASITES.

So the male's display is just a way to show off how healthy and clever he is.

CORRECT.

MY GOODNESS, WHAT A LOT OF BOTHER TO GO THROUGH JUST TO REPRODUCE. GIVE ME A NICE LITTLE BUD FORMING ON MY BACK ANY DAY OF THE WEEK.

IT **IS** STRANGE, YOUR MOST RESPLENDENT ROYAL REPRODUCER, BUT THESE TYPES OF DISPLAYS PLAY A CRUCIAL ROLE IN THE EVOLUTION OF NEW SPECIES BY KEEPING TWO GROUPS OF ORGANISMS SEPARATED. HERE'S AN EXAMPLE.

ON THE ISLAND OF **IRIAN JAYA** IN INDONESIA, GEOGRAPHIC ISOLATION AND SEXUAL SELECTION PROVIDE A ONE-TWO SPECIATION PUNCH FOR TWO POPULATIONS OF VOGELKOP BOWERBIRDS.

IRIAN JAYA

SEXUAL SELECTION

GEO-GRAPHIC SELECTION

UH... I'M A LOVER, NOT A FIGHTER.

THESE TWO POPULATIONS APPEAR TO BE THE SAME SPECIES. THEY ARE NEARLY INDISTINGUISHABLE PHYSICALLY AND GENETICALLY. IN FACT, ABOUT THE ONLY THING DIFFERENT BETWEEN THE TWO POPULATIONS IS THAT THEY BUILD DIFFERENT BOWERS.

ARFAK

FAKFAK

IRIAN JAYA

THE POPULATION LIVING IN THE **ARFAK MOUNTAINS** BUILD ELABORATE HUTS THAT THEY DECORATE WITH BRIGHT BLUE AND RED FLOWERS.

ARFAK

THE MALES OF THE POPULATION LIVING IN THE **FAKFAK MOUNTAINS** BUILD BOWERS THAT LOOK LIKE MAYPOLES AND ARE ADORNED WITH DRAB BEIGE AND BROWN DECORATIONS.

FAKFAK

GEOGRAPHIC ISOLATION MADE IT POSSIBLE FOR THESE DIFFERENT DISPLAYS TO EVOLVE INDEPENDENTLY, BUT SEXUAL SELECTION **REINFORCES** THE DIFFERENCE.

A FEMALE FAKFAK BOWERBIRD WON'T MATE WITH A MALE FROM THE ARFAK POPULATION NO MATTER HOW SIMILAR THEY ARE GENETICALLY BECAUSE HE ISN'T SENDING THE RIGHT SIGNALS.

YOU KNOW, YOU'RE GOOD-LOOKING, BUT I JUST DON'T UNDERSTAND YOU.

SOMETHING NEW UNDER THE SUNFLOWER

SUNFLOWER TRANSFORMATIONS HAVE MYTHICAL ROOTS. THE ANCIENT GREEKS TELL US THAT WHEN THE NYMPH CLYTIE WAS REJECTED BY THE SUN GOD APOLLO, SHE SAT WATCHING HIM CROSS THE SKY FOR DAYS WITHOUT EATING OR DRINKING A THING. AFTER SEVERAL DAYS OF STARING, SHE SLOWLY METAMORPHOSED INTO THE SUN-LOVING SUNFLOWER. IT'S A PRETTY GOOD STORY. BUT BIOLOGIST LOREN RIESEBERG AND HIS COLLEAGUES AT INDIANA UNIVERSITY HAVE A FAR LESS HEARTBREAKING STORY OF SUNFLOWER ORIGINS THAT INVOLVES THE FORMATION OF A HYBRID.

HYBRIDS ARE THE OFFSPRING PRODUCED WHEN TWO CLOSELY RELATED SPECIES MATE AND PRODUCE YOUNG. USUALLY, HYBRIDS ARE STERILE AND LESS SUCCESSFUL AT SURVIVING THAN PUREBRED OFFSPRING. SOMETIMES, HOWEVER, FERTILE HYBRIDS ARE FORMED THAT HAVE REMARKABLE ADAPTATIONS NOT PRESENT IN THE PARENT SPECIES. SUCH IS THE CASE WITH SUNFLOWERS.

LOREN RIESEBERG'S GROUP HAS FOUND THAT SUNFLOWERS HAVE GENERATED IMPRESSIVELY ADAPTIVE HYBRIDS FROM THE "PARENT" SUNFLOWER SPECIES *HELIANTHUS ANNUUS* AND *HELIANTHUS PETIOLARIS*. THE HYBRID SPECIES ARE CAPABLE OF EXPLOITING EXTREME ENVIRONMENTS THAT THE PARENT SPECIES CANNOT. *HELIANTHUS ANOMALUS* THRIVES IN SAND DUNES, *HELIANTHUS DESERTICOLA* GROWS IN DRY, SANDY SOIL IN THE DESERT, AND *HELIANTHUS PARADOXUS* LIVES IN BRACKISH SALTWATER MARSHES.

THE GENETIC MAKEUP OF THESE THREE HYBRIDS INDICATES THAT THEY EVOLVED PRETTY RECENTLY, WHICH MEANS THAT THE PARENT SPECIES HAVEN'T HAD MUCH TIME TO EVOLVE SINCE THEY PRODUCED THE HYBRIDS. RIESEBERG'S TEAM USED THIS FACT TO DO AN ELEGANT EVOLUTIONARY EXPERIMENT. THEY MATED THE PARENT SPECIES AND GOT OFFSPRING SIMILAR TO THE EXTREME HYBRIDS GENERATED THOUSANDS OF YEARS AGO. JUST LIKE THE ANCIENT HYBRIDS, THESE NEW HYBRIDS COULD LIVE IN PLACES THEIR PARENTS COULD NOT.

THESE RESULTS DEMONSTRATE THE POTENTIALLY POWERFUL ROLE HYBRIDS COULD PLAY IN EVOLUTION. A SINGLE GENE MUTATION IN A SUNFLOWER WOULDN'T BE ENOUGH TO GENERATE THE PHENOTYPIC VARIATION NEEDED TO MOVE INTO A RADICALLY DIFFERENT ENVIRONMENT.

BUT HYBRIDS CAN PROVIDE VARIATION AT HUNDREDS, MAYBE THOUSANDS, OF DIFFERENT GENES. THIS LEVEL OF VARIATION MADE IT POSSIBLE FOR SUNFLOWER HYBRIDS TO MAKE MAJOR ECOLOGICAL TRANSITIONS IN A SINGLE GENERATION.

That's like when the banded orbi danced for the woolly orbi.

THAT'S RIGHT. SEXUAL SELECTION CAN DRIVE THE EVOLUTION OF PHYSICAL AND BEHAVIORAL TRAITS THAT ISOLATE RELATED SPECIES FROM EACH OTHER AND PREVENT THEM FROM MATING.

So what happens if, after **ALL** of that, a couple of organisms still make a mistake and choose a mate from a different species?

I'M A PRINCE, **REALLY**. KISS ME AND SEE.

OKAY. MAYBE WE CAN MAKE THIS WORK.

THE OFFSPRING OF TWO DIFFERENT BUT RELATED SPECIES IS CALLED A **HYBRID**.

IN SOME PLANTS, LIKE SUNFLOWERS, HYBRIDS ARE BETTER SUITED FOR CERTAIN ENVIRONMENTS AND FORM NEW SPECIES.

ONE SMALL STEP FOR A PLANT, ONE GIANT LEAP FOR PLANTKIND.

BUT MOST OF THE TIME, THE HYBRID STORY ISN'T A HAPPY ONE. HYBRIDS ARE USUALLY EITHER **NON-VIABLE** OR **STERILE**. IN NONVIABLE HYBRIDS, THE DNA INSTRUCTIONS FROM THE EGG AND SPERM ARE TOO DIFFERENT, AND THE ZYGOTE NEVER DEVELOPS INTO A BABY.

ZYGOTE

I GIVE UP.

I CAN'T MAKE HEADS OR TAILS OF THESE INSTRUCTIONS.

DNA USER'S GUIDE

HYBRID STERILITY OCCURS WHEN A HYBRID ZYGOTE SUCCESSFULLY GROWS INTO AN ADULT BUT THAT ADULT CANNOT PRODUCE OFFSPRING. FOR EXAMPLE, HORSES AND DONKEYS ARE SEPARATE SPECIES, BUT EVERY ONCE IN A WHILE THEY WILL MATE. ON THOSE OCCASIONS THEY PRODUCE A HYBRID CALLED A **MULE**.

SMILE, DEAR, OUR PICTURE IS GOING TO BE IN A BOOK.

NO.

SIGH. HE'S SO STUBBORN.

MULES ARE PERFECTLY VIABLE CREATURES EXCEPT FOR THE FACT THAT THEY CANNOT PRODUCE OFF-SPRING.

SO THAT'S THE BARE BONES OF **SPECIATION**. KEEP IN MIND THAT EARTH SCIENTISTS HAVE WRITTEN VOLUMES ON THE TOPIC. THIS IS INTENDED ONLY AS A WAY TO INTRODUCE THE BASICS.

Well, let's see if I understand those basics, then.

DOOT

SPECIATION

1. Speciation is the process in which one or two new species evolve from a single ancestral species.

2. It can occur when two populations of the same species are separated from each other in some way.

3. When this happens, each population will follow its own evolutionary path, accumulating its own set of random mutations in response to environmental pressure.

4. Over time, the two populations acquire so many differences that they cannot or will not mate with each other anymore.

5. If they make a mistake and do mate with the wrong species, the hybrid offspring produced is often nonviable and fails to develop.

6. If it does develop, it will either be sterile or less fit than its parent species and so less likely to survive and reproduce.

On rare occasions, the hybrid is successful and can actually become a new species.

YOUR MEMORY IS SUR-PASSED ONLY BY YOUR UNPARALLELED SKILL AT STATING THINGS SUCCINCTLY, YOUR HIGHNESS.

SO IT SEEMS LIKE OUR OPTIONS IN COMBATING THIS GENETIC PLAGUE AFFLICTING OUR SPECIES ARE TO EITHER EVOLVE **SEXUAL REPRODUCTION** AND MIX GENES...

EWWWW

...OR GENERATE SQUINCH HYBRIDS THAT CONTAIN GENES FROM OTHER SPECIES THAT MIGHT CONFER SOME KIND OF ADVANTAGE.

I vote for option #2.

Besides, evolving sexual reproduction couldn't happen quickly, and certainly not in my lifetime. But is the second option even possible?

IT MAY BE, YOUR HIGHNESS. HUMANS HAVE A LONG HISTORY OF BREEDING ORGANISMS FOR FAVORABLE TRAITS.

Really?

YES, AND THEY HAVE ALSO DEVELOPED THE TECHNOLOGY TO INSERT GENES FROM ONE ORGANISM INTO ANOTHER.

CLEARLY THERE IS STILL **MUCH** WE NEED TO CONSIDER.

MIGHT I SUGGEST WE BEGIN BY EXPLORING THE HOST OF **ADAPTATIONS** THAT DIFFERENT SPECIES HAVE EVOLVED TO SURVIVE IN VARIOUS CONDITIONS?

Why do I get the feeling that that is the next stop on the tour?

ISN'T THAT A HAPPY COINCIDENCE, SIRE?

CHAPTER 5

Perfectly Imperfect

So the humans have found a way to shape adaptations of other species to their liking?

INDEED, YOUR HIGHNESS.

HUMANS HAVE INTUITIVELY USED THE PRINCIPLES OF NATURAL SELECTION TO BREED DOMESTICATED PLANTS AND ANIMALS FOR NEARLY 10,000 YEARS. CHARLES DARWIN CALLED THE PROCESS **ARTIFICIAL SELECTION**.

ARTIFICIAL? Is it done by **ROBOTS** or something?

NO, NO. DARWIN CALLED IT THAT BECAUSE **HUMANS** DID THE SELECTING INSTEAD OF **NATURE**.

PREHISTORIC FARMERS RECOGNIZED VARIATION IN THEIR LIVESTOCK. THEY UNDERSTOOD THAT IF THEY CONTROLLED WHICH PLANTS AND ANIMALS WERE ALLOWED TO BREED, THEY COULD EVENTUALLY CHANGE THE PHYSICAL AND BEHAVIORAL CHARAC-TERISTICS OF THEIR CROPS AND LIVESTOCK.

I'M AFRAID I HAVE TO LET YOU GO.

BY SELECTING FOR DIFFERENT PARTS OF THE PLANT, HUMANS HAVE BRED BRUSSELS SPROUTS, KALE, CAULIFLOWER, TURNIPS, RUTABAGA, AND KOHLRABI-- ALL FROM A SINGLE, **PRIMITIVE CABBAGE** SPECIES.

THEY'RE ALL DERIVED FROM THE SAME PLANT, DEAR.

AND I WON'T EAT **ANY** OF THEM.

USING ARTIFICIAL SELECTION, HUMANS HAVE BRED CHICKENS FROM RED JUNGLE FOWL...

...PIGS FROM WILD BOARS...

...AND COWS FROM AUROCHS.

THEIR EFFORTS HAVE EVEN TRANSFORMED FIERCE PREDATORS LIKE WOLVES...

B-B-BAD DOGGIE.

GRRRR

...INTO DOCILE DOGGY COMPANIONS.

GOOD DOGGY.

Humans must be astoundingly patient. Surely it took thousands of years for such a transformation.

ACTUALLY, DMITRI BELYAEV'S WORK WITH WILD SILVER FOXES DEMONSTRATED THAT IT COULD HAPPEN QUITE QUICKLY.

IT WAS NOT SO HARD, REALLY.

ALTHOUGH WILD FOXES ARE QUITE AGGRESSIVE TOWARD HUMANS, WE DRAMATICALLY ALTERED THEIR BEHAVIOR WITHIN FIFTY YEARS. OUR EXPERIMENTAL PROCEDURE WAS STRAIGHTFORWARD. WE SIMPLY SELECTED THE MOST DOCILE FOXES TO BREED IN EACH GENERATION.

THIS ONE, DA.

THAT ONE, NYET.

"WITHIN EIGHTEEN GENERATIONS WE HAD BRED A POPULATION OF FRIENDLY FOXES WITH FLOPPY EARS AND PATCHY COAT COLORATION. THEY WAGGED THEIR TAILS AND BARKED. THEY EVEN SHOWED AFFECTION BY LICKING US, JUST LIKE A DOG. WILD FOXES DO NONE OF THESE THINGS."

STOP THAT! HAVE SOME DIGNITY, FOR PETE'S SAKE!

18 GENERATIONS

Good grief! How could you get all of those changes at once?

THE ANSWER LIES IN THE FOX'S **GENES**, YOUR HIGHNESS.

SELECTING FOR A SPECIFIC PHENOTYPIC TRAIT MEANS YOU ARE ALSO SELECTING FOR THE GENE OR GENES THAT MAKE THE TRAIT.

IN THIS CASE, BY SELECTING FOR TAMENESS, IT TURNED OUT WE WERE CHOOSING FOXES WITH LOWER LEVELS OF A HORMONE CALLED **ADRENALINE**, WHICH IS USED IN AGGRESSIVE BEHAVIOR.

I still don't see how **TAMENESS** and **COAT COLOR** are related.

Unless this adrenaline stuff is for something more than just aggressive behavior...

DA -- THAT WAS THE ANSWER!

NOT ONLY IS ADRENALINE USED IN FOX AGGRESSION, IT IS ALSO NEEDED TO MAKE THE PIGMENTS IN A FOX'S SKIN AND FUR. SINCE TAME FOXES HAVE LOW LEVELS OF ADRENALINE, THEY ALSO HAVE LOW LEVELS OF THE PIGMENT THAT PRODUCES COLOR IN THEIR FUR.

ADRENALINE LOW

ADRENALINE HIGH

HEY, LET'S GO FETCH SOMETHING.

YOU ARE AN EMBARRASSMENT TO FOXES EVERYWHERE.

So changing one trait had several unexpected effects on other traits.

PRECISELY, YOUR MOST PERSPICACIOUS PRINCELINESS.

THIS IS CALLED A **PLEIOTROPIC EFFECT.** PLEIOTROPIC CHANGES OCCUR WHEN A SINGLE GENE, SUCH AS THE GENE FOR MAKING ADRENALINE, INFLUENCES MULTIPLE, SEEMINGLY UNRELATED PHYSICAL TRAITS, LIKE TAMENESS AND COAT COLOR.

Are they **BOTH** adaptations, then?

AN EXCELLENT QUESTION, YOUNG HIGHNESS. AT THE START OF THIS TOUR I SAID AN ADAPTATION WAS A "SPECIALIZED FEATURE," BUT NOW WE NEED TO REFINE THAT A BIT.

AN **ADAPTATION** IS ANY FEATURE OF AN ORGANISM THAT GIVES IT AN ADVANTAGE IN THE STRUGGLE TO SURVIVE AND REPRODUCE.

CRACK!

SO, I PUT IT TO YOU, YOUR HIGHNESS: WHICH FEATURE OF THE FOXES HELPED THEM SURVIVE UNDER HUMAN SELECTION? TAMENESS OR COAT COLOR?

Well, I guess **TAMENESS.**

Coat color was just a **SIDE EFFECT.**

I CONCUR WITH YOUR ASSESSMENT. BELYAEV'S WORK HIGHLIGHTS TWO IMPORTANT POINTS. FIRST, PHENOTYPIC CHANGES CAN OCCUR QUICKLY -- EVEN DRAMATIC ONES -- AND INVOLVE ONLY A FEW GENES.

SECOND, BELYAEV DEMONSTRATED THAT SELECTION FOR A SINGLE TRAIT CAN GENERATE UNEXPECTED **VARIATION** IN A POPULATION.

And more variation means more **POSSIBILITIES**, right, Bloort?

INDEED. VARIATION CAN LEAD TO BRAND-NEW EVOLUTIONARY INNOVATIONS. BUT IN SOME CASES, IT CAN ALSO LEAD TO NEW FUNCTIONS FOR EXISTING TRAITS.

What do you mean, "new functions for existing traits"? Are you saying a trait can just **SHIFT** what it does?

IT'S QUITE COMMON UNDER THE RIGHT CONDITIONS, YOUR HIGHNESS.

A **FUNCTIONAL SHIFT** OCCURS WHEN A TRAIT ADAPTED FOR ONE FUNCTION STARTS TO BE USED FOR SOMETHING COMPLETELY DIFFERENT.

I still don't see how that could happen.

LET'S LOOK FOR AN EXAMPLE HERE IN EARTH'S AMAZON RIVER.

BOOP.

I can't look for anything in here. This murky water makes it impossible to see.

FOR A SQUINCH EYE, PERHAPS, SIRE. BUT NAVIGATING THESE WATERS IS NO PROBLEM FOR THE WEAKLY ELECTRIC **ELEPHANT FISH**.

Really? How does a fish so **DUMB** find its way around this place?

I NEVER SAID IT WAS DUMB, SIRE.

You said it was **WEAKLY ELECTRIC**.

I just assumed you meant it was "dim."

giggle

D-a-a-a-d, you're **EMBARRASSING** me.

NO, NO -- A TRULY **DELIGHTFUL** JEST, O MONARCH OF MIRTH. IN ACTUALITY, THESE FISH GENERATE AN ELECTRICAL FIELD AROUND THEIR BODIES THAT THEY USE TO SENSE OBSTACLES IN THE ENVIRONMENT.

DEADWOOD OFF THE STARBOARD BOW. DIVE! DIVE!

AROOGA! AROOOGA!

YOU ARE SO WEIRD.

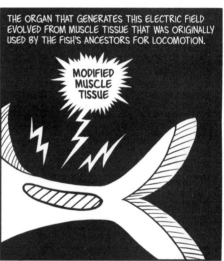

THE ORGAN THAT GENERATES THIS ELECTRIC FIELD EVOLVED FROM MUSCLE TISSUE THAT WAS ORIGINALLY USED BY THE FISH'S ANCESTORS FOR LOCOMOTION.

MODIFIED MUSCLE TISSUE

So, generating an electric field had a greater survival value for the elephant fish than having an extra tail muscle?

EXACTLY, MY MOST PERCEPTIVE POTENTATE. AND IT GETS BETTER. THE AMAZON IS ALSO HOME TO THE **ELECTRIC EEL.**

ELECTRIC EELS EVOLVED FROM WEAKLY ELECTRIC FISH SIMILAR TO THE ELEPHANT FISH. BUT THE EEL'S ELECTRIC ORGAN HAS FURTHER EVOLVED TO DIS-CHARGE A DEADLY ELECTRIC SHOCK OF **500 VOLTS.**

HERTZ, DON'T IT?

ACK. WHAT A RE-**VOLT**-ING DEVELOPMENT.

So...the function of this fish muscle tissue has shifted from **LOCOMOTION**...

...to **SENSATION**...

...to **LETHAL WEAPON**.

I see. New adaptations can arise when new uses for existing traits have a higher survival value for the organism.

That's very cool, Bloort, but...

IS THERE SOMETHING TROUBLING MY FUTURE LORD AND MASTER?

Well, yeah, I was just wondering...based on your definition, how do we know that all these things you're showing us are **ADAPTATIONS**?

Oh, come now, Son. Surely the ability to electrocute your prey is an obvious selective advantage.

It certainly **SEEMS** so, but if we're looking for adaptations to introduce into **OUR** gene pool, shouldn't we have some kind of scientific **EVIDENCE** that it **IS** an **ADVANTAGE**?

I mean, otherwise it's just a good story, right, Bloort?

YOUR HIGHNESS, THERE ARE TIMES IN OUR LIVES WHEN WE GET TO WITNESS A TRULY ASTUTE MIND PEELING BACK THE LAYERS OF THE UNIVERSE TO FIND A DAZZLING NUGGET OF WISDOM.

I AM DELIGHTED TO NOW BEAR WITNESS TO SUCH AN EVENT.

INDEED, EVOLUTIONARY BIOLOGISTS ARE OFTEN CONFRONTED WITH AMAZING TRAITS. BUT THEY MUST STILL PRODUCE EVIDENCE THAT THESE ARE ADAPTATIONS.

CONSIDER THE CURIOUS CASE OF THE MARINE ISOPOD PARACERCEIS SCULPTA.

THIS CREATURE LIVES IN THE OCEAN AND REPRODUCES INSIDE OF SPONGES. THIS IS WHAT THE FEMALE LOOKS LIKE.

FEMALE
PARACERCEIS SCULPTA

AND THIS IS WHAT THE MALES LOOK LIKE.

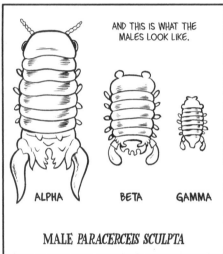

ALPHA BETA GAMMA

MALE *PARACERCEIS SCULPTA*

Which one is the male?

THEY ALL ARE, SIRE.

Uh...isn't that middle one another female?

NO, EACH OF THESE IS A RE-PRODUCTIVELY MATURE MALE. THIS SPECIES HAS **THREE DIFFERENT** MALE PHENO-TYPES. THE **ALPHA** MALE IS BIG AND BURLY, THE **BETA** MALE IS MEDIUM-SIZED AND LOOKS LIKE A FEMALE, AND THE **GAMMA** MALE IS SMALL.

THE ALPHAS USUALLY SET UP SHOP IN A SPONGE AND GUARD A HAREM OF FEMALES INSIDE, KICKING OUT ANY OTHER MALES THAT TRY TO GET IN. BETAS AND GAMMAS ARE PHYSICALLY NO MATCH FOR THE ALPHA. AND YET THEY CAN BE FOUND INSIDE THE SPONGE, TRYING TO MATE WITH THE FEMALES.

Why doesn't the alpha male kick them out?

HE DOESN'T KNOW THEY'RE THERE.

THE GAMMA HIDES FROM THE ALPHA IN THE LITTLE NOOKS AND CRANNIES OF THE SPONGE. HE ATTEMPTS TO MATE WITH THE FEMALES ONLY WHEN THE COAST IS CLEAR.

?....

THE BETA, HOWEVER, IS TOO BIG TO HIDE, SO WHEN THE ALPHA COMES AROUND THE BETA STARTS ACTING LIKE A FEMALE. THE ALPHA TRIES TO COURT HIM, BUT WITH NO SUCCESS.

AMAZINGLY ENOUGH, THE PHYSICAL AND BEHAVIORAL DIFFERENCES AMONG THE THREE MALE PHENOTYPES ARE THE RESULT OF VARIATIONS IN A SINGLE GENE.

By the Seven Nebulas of Maagoo, Bloort, Earth is soooo weird.

So, if each phenotype is the result of variations in a single gene, which form is most adaptive?

Yes. Which one is **BEST**, Bloort?

YOU TELL ME.

I beg your pardon? Did you just give your King...

...a **COMMAND?**

UH--

NO! YES! I MEAN, I DIDN'T MEAN... I **BEG YOUR PARDON** FOR MY INFOR- MAL QUESTION, SIRE.

I ONLY MEANT THAT I BELIEVE THE **PRINCE** HAS THE INTELLECTUAL ACUMEN TO IMAGINE A WAY OF TESTING WHICH, IF ANY, OF THESE PHENOTYPES ARE ADAPTIVE.

97

Hmm. We could figure out which one of the phenotypes leaves more offspring...

BRILLIANT! THAT IS PRECISELY WHAT STEPHEN M. SHUSTER AND HIS COLLEAGUES DID.

THEY DID LABORATORY MATING EXPERIMENTS AND ALSO SURVEYED THE MALE PHENOTYPES FOUND IN WILD SPONGES. LOOKING AT THOSE RESULTS, THEY DETERMINED HOW MANY OFFSPRING EACH OF THE DIFFERENT MALES PRODUCED.

And?

THEY'RE ALL EQUAL. EACH PHENOTYPE IS ADAPTIVE IN ITS OWN WAY.

THE ALPHA CAN FIGHT OFF ALPHA COMPETITORS TO MAINTAIN HIS HAREM.

POW

THE BETAS CAN DECEIVE THE ALPHA AND GAIN ACCESS TO THE FEMALES THAT WAY.

THE GAMMAS CAN SNEAK MATING OPPORTUNITIES BY HIDING IN THE SPONGE.

IN NATURE, EACH STRATEGY IS EQUALLY EFFECTIVE IN MAKING BABIES.

Ha-ha! That's incredible.

Pfft. Incredibly **BIZARRE**, if you ask me.

OH, ADAPTATIONS COME FAR STRANGER THAN THAT, YOUR HIGHNESS. JUST IMAGINE IF YOU WERE A FLOUNDER AND HAD TO DEAL WITH...

... THE **MIGRATING EYE!**

FLOUNDER ARE SEA FISH THAT MAKE THEIR LIVING LYING ON THEIR SIDES ON THE OCEAN FLOOR AND AMBUSHING THEIR PREY.

OH, JIMMY, I'M SCARED. I FEEL LIKE I'M BEING **WATCHED**.

AH, DON'T TELL ME YOU BELIEVE THOSE STORIES ABOUT OLD MAN FLOUNDER RISING FROM THE SEA FLOOR TO CLAIM HIS VICTIMS?

BWAH HA-HA-HAA ...

WHEN IT'S YOUNG, A FLOUNDER LOOKS AND ACTS LIKE ANY OTHER FISH SWIMMING IN THE SEA.

BUT AS IT BEGINS TO TURN INTO AN ADULT, IT UNDERGOES A STRANGE METAMORPHOSIS. THIRTY TO FORTY DAYS AFTER HATCHING, ITS RIGHT EYE AND RIGHT NOSTRIL BEGIN TO MIGRATE OVER TO THE LEFT SIDE OF ITS HEAD.

OH, NO! THE SCHOOL DANCE IS TONIGHT.

HA!

LOOK AT THAT! HA HA!

HA! HA!

Parasites on the Brain

by Per A. Sidik

The world is filled with organisms adapted to exploit other creatures. Parasites live in or on another organism -- called a host -- and use it for survival. This lifestyle has driven the evolution of several specialized adaptations to help the parasite find a host, avoid being digested or attacked by the host's immune system, and avoid killing the host before reproducing.

Some of these adaptations are molecules on the parasites' bodies or eggs that trick the host's immune system into thinking the parasite isn't a threat.

In some cases, parasites have adapted to avoid detection by eating the organs responsible for the host's immune system, while other parasites release chemicals that turn off the host's immune system altogether. Once the parasite has reproduced, its eggs leave the host, usually exiting with the poop, so that the offspring can infect someone else.

But how do the eggs and young parasites find another host? Some have adapted to release chemicals called hormones and neurotransmitters to control another organism's brain. The eggs of a parasitic worm called a **NEMATODE** begin life in the abdomens of common ants. The eggs, however, won't hatch until they enter the body of a bird.

How do they make the insect-to-avian leap? First, the nematode eggs turn the black ant's abdomen bright red, making it look like a delicious berry treat. Second, the eggs release chemicals that stimulate the ant to carry its red backside high in the air, making it an easy target for a hungry bird.

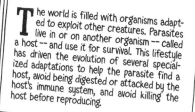

Death by berry-butt is a bad way to go, but perhaps not as bad as being the undead slave to a parasite's children. A few parasitic wasps have mastered the art of making **ZOMBIE SLAVES** by laying their eggs in caterpillars. A few weeks later, most of the wasp larvae burst out of the host, but a few larvae stay behind in the caterpillar, releasing chemicals that dramatically alter the half-dead caterpillar's behavior. The formerly mild-mannered caterpillar now stands guard over the other young wasps, aggressively head-butting and swatting away intruders as the wasps complete their metamorphosis. This zombie nanny has been shown in experiments to dramatically increase the survival rate of the young wasps. (And, best of all, it charges only four dollars an hour!)

ONCE THE EYE AND NOSTRIL MIGRATION IS COMPLETE, THE FLOUNDER SETTLES TO THE OCEAN FLOOR AND AWAITS ITS FIRST VICTIM.

THEY LAUGHED AT ME BEFORE, BUT I'LL SHOW THEM. I'LL SHOW THEM ALL!

Ack, that makes my eye ache just thinking about it.

How does it happen?

WELL...

...THE DEVELOPMENT OF THE SKULL HAS TO BE ALTERED DRAMATICALLY. HEAD BONES CONTORT, SOME BONES DISAPPEAR, AND OTHERS GROW RAPIDLY TO PUSH THE EYE SOCKET AROUND.

Isn't that... I don't know... a bit...

ELABORATE?

I was going to say "over the top."

giggle

Ha-ha! Excellent sense of humor. Like father, like son!

HIGH FIVE!

HUMOROUS AND ACCURATE, SIRE. TO UNDERSTAND THE MIGRATING EYE, WE MUST REMEMBER THAT NATURAL SELECTION CAN WORK ONLY WITH WHAT IT'S GOT. MUTATIONS ARE RANDOM, NOT PLANNED, SO ADAPTATIONS MUST BE MADE FROM THE MATERIALS ON HAND.

FLOUNDER MUTATIONS

10-YEAR PLAN

CONSEQUENTLY, MANY ADAPTATIONS LOOK LIKE CRAZY CONTRAPTIONS. TO ILLUSTRATE THIS POINT FURTHER, ALLOW ME TO GIVE YOU SOME FLOWERS.

Oh, thank you, Bloort. They're **LOVELY**.

ACTUALLY, THEY'RE REPRODUCTIVE STRUCTURES. FLOWERS ARE THE SOURCE OF THE MALE POLLEN AND FEMALE OVULES THAT ANGIOSPERMS USE TO PROCREATE.

Yuck, yuck, and yuck.

FLOWERS REPRODUCE WHEN INSECTS CARRY POLLEN FROM ONE FLOWER TO A DIFFERENT FLOWER OF THE SAME SPECIES. THIS POLLEN USUALLY GETS STUCK IN THE INSECT'S HAIRS WHEN IT IS RUMMAGING AROUND THE FLOWERS, LOOKING FOR NECTAR.

So, insects are the flowers' unintentional matchmakers?

IN A SENSE, BUT MOST OF THE TIME THEY DO GET A SUGARY TREAT FOR THEIR TROUBLE.

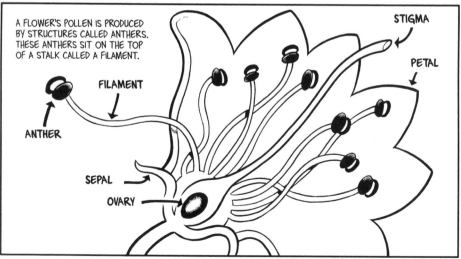

A FLOWER'S POLLEN IS PRODUCED BY STRUCTURES CALLED ANTHERS. THESE ANTHERS SIT ON THE TOP OF A STALK CALLED A FILAMENT.

STIGMA

PETAL

FILAMENT

ANTHER

SEPAL

OVARY

IN MOST FLOWERS, ANTHERS JUST SIT IN THERE, WAITING FOR AN INSECT TO BRUSH UP AGAINST THEM. BUT IN THE **BUNCH-BERRY** FLOWER, THE ANTHER AND FILAMENT ARE PART OF AN ELABORATE POLLEN **BOOBY TRAP.**

THE PETALS OF THE BUNCH-BERRY FLOWER ARE HELD SHUT OVER THE ANTHERS, CAUSING THE FILAMENT TO BEND. BECAUSE OF THIS, THE ANTHERS ARE HELD UNDER A LOT OF TENSION AND ARE READY TO SWING UP WHEN THE PETALS OPEN.

BUNCHBERRY ESCAPE PLAN

FILAMENT
PETAL
ANTHER

① SPRING-LOADED ANTHERS AND HOLD BACK WITH PETALS.

② WAIT FOR UNSUSPECTING POLLEN-SPREADER TO TRIP THE PETAL.

③

POLLEN

WHEN A LARGE INSECT LIKE A BUMBLEBEE VISITS THE BUNCHBERRY FLOWER...

DUM DEED UM DUM

...IT TRIPS A TRIGGER ON ONE OF THE PETALS, CAUSING THE FOUR PETALS TO EXPLODE OPEN IN LESS THAN .0005 SECONDS...

WHAT THE -- ?

...RELEASING THE ANTHERS, WHICH FLIP UP LIKE A MEDIEVAL CATAPULT, FLINGING THE POLLEN UP INTO THE INSECT'S BODY HAIRS.

OH, THIS IS GOING TO BE **GREAT** FOR MY ALLERGIES.

Such a forceful ejection would ensure that some of the pollen gets stuck in the bee's hair.

Ingenious.

ESPECIALLY SINCE THIS REMARKABLE APPARATUS WASN'T DESIGNED AND BUILT FROM SCRATCH WITH BRAND-NEW CATAPULT STRUCTURES PERFECTLY SUITED TO THE TASK. IT EVOLVED FROM EXISTING PETALS AND ANTHERS.

Yes, yes, it is all very remarkable. But those petals are so tiny and bland, not big, floppy, and pretty like the flowers you gave me, Bloort.

I APOLOGIZE, MY LIEGE, BUT NATURAL SELEC-TION'S TINKERING COMES AT A **PRICE.**

THE PETALS CANNOT BE BOTH FLOPPY **AND** A TAUT, SPRING-LOADED POLLEN TRAP. THIS IS WHAT WE CALL AN EVOLUTIONARY **TRADE-OFF**.

You're starting to sound like one of my economic advisors, Bloort.

THAT MAY BE BECAUSE THERE ARE COSTS AND BENEFITS TO ALL ADAPTATIONS.

A beneficial adaptation can have a **DOWNSIDE**?

ABSOLUTELY, YOUR HIGHNESS. A **TRADE-OFF** OCCURS WHEN AN ADAPTATION PROVIDES A BENEFIT BUT ALSO COMES WITH A COST. WE CAN SEE THIS QUITE CLEARLY IN THE INTERACTION OF **MOTHS** AND **BOLAS SPIDERS**.

WHEN THE FEMALE MOTH IS RECEPTIVE TO MATING, SHE ALERTS NEARBY MALES BY RELEASING A CHEMICAL CALLED A PHEROMONE INTO THE AIR.

RRRROW!! WHAT IS THAT PERFUME?

DO YOU LIKE IT? I MADE IT MYSELF.

MALES PICK UP THE SCENT WITH THEIR ELABORATELY BRANCHED ANTENNAE AND GO LOOKING FOR LOVE.

AH, LOVE IS IN THE AIR.

TO AVOID CONFUSION, FEMALES OF EACH SPECIES OF MOTH PRODUCE A PERFUME THAT IS ATTRACTIVE ONLY TO THE MALES OF THE SAME SPECIES.

WHAT DOES HE SEE IN **HER**?

Sounds like a pretty good system. What's the downside?

THE PREVIOUSLY MENTIONED BOLAS SPIDERS.

MOST SPIDERS EAT A VARIETY OF PREY, BUT BOLAS SPIDERS EAT ONLY MALE MOTHS.

See? I'm not the only picky eater in the galaxy.

Yes, yes, Son. But what makes these spiders so persnickety, Bloort?

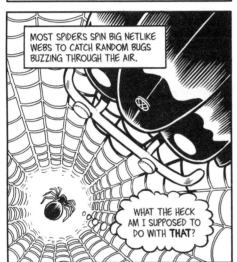

MOST SPIDERS SPIN BIG NETLIKE WEBS TO CATCH RANDOM BUGS BUZZING THROUGH THE AIR.

WHAT THE HECK AM I SUPPOSED TO DO WITH **THAT**?

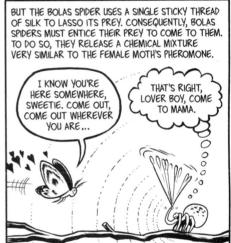

BUT THE BOLAS SPIDER USES A SINGLE STICKY THREAD OF SILK TO LASSO ITS PREY. CONSEQUENTLY, BOLAS SPIDERS MUST ENTICE THEIR PREY TO COME TO THEM. TO DO SO, THEY RELEASE A CHEMICAL MIXTURE VERY SIMILAR TO THE FEMALE MOTH'S PHEROMONE.

I KNOW YOU'RE HERE SOMEWHERE, SWEETIE. COME OUT, COME OUT WHEREVER YOU ARE...

THAT'S RIGHT, LOVER BOY, COME TO MAMA.

It's a mimic, just like that bowerbird you talked about earlier.

INDEED, YOUR HIGHNESS. BUT INSTEAD OF WOOING A **MATE**, THE BOLAS SPIDER IS LURING A **MEAL**.

WHEN THE UNSUSPECTING MALE APPROACHES, THE BOLAS SPIDER HITS HIM WITH THE STICKY STRING AND REELS HIM IN.

YOU TRICKED ME!

OF COURSE. I'M AN AMAZING SPIDER, MAN.

THWP!

Good gracious, that seems pretty unsporting.

ALL'S FAIR IN THE PURSUIT OF LOVE AND LUNCH, YOUR HIGHNESS. THE MOTHS ARE USING A VERY SPECIFIC CODE, AND THE BOLAS SPIDER HAS CRACKED IT.

Then why do the moths persist in using pheromones? Surely this trait would be selected against and soon disappear.

NOT NECESSARILY, YOUR HIGHNESS. THE REPRODUCTIVE BENEFITS OF USING PHEROMONES TO COMMUNICATE OUTWEIGH THE COST.

A FEW MALE MOTHS ARE CAUGHT BY BOLAS SPIDERS, BUT THERE ARE MANY MORE MOTHS THAN SPIDERS. THE VAST MAJORITY AVOID GETTING EATEN AND HAVE A SHOT AT REPRODUCING.

Hmm, yes, I see. Just like in economics, evolutionary choices can sometimes have nasty consequences.

YES, YOUR HIGHNESS. SOMETIMES EVOLUTION CAN PAINT A SPECIES INTO A CORNER, CREATING AN EVOLUTIONARY **CONSTRAINT**. A "CONSTRAINT" IS AN ADAPTATION THAT LIMITS A SPECIES' FUNCTIONAL AND EVOLUTIONARY POSSIBILITIES.

Fascinating. I don't suppose you have an example of that?

INDEED I DO, YOUR HIGHNESS. BEHOLD, THE **ICEFISH** OF **ANTARCTICA**.

BRRR

This part of the tour is a bit too realistic, Bloort.

NOTED, SIRE. THIS IS AN ICEFISH.

Whoa! You can see right through it.

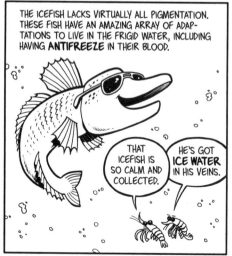

THE ICEFISH LACKS VIRTUALLY ALL PIGMENTATION. THESE FISH HAVE AN AMAZING ARRAY OF ADAPTATIONS TO LIVE IN THE FRIGID WATER, INCLUDING HAVING **ANTIFREEZE** IN THEIR BLOOD.

THAT ICEFISH IS SO CALM AND COLLECTED.

HE'S GOT **ICE WATER** IN HIS VEINS.

BUT PERHAPS THE MOST AMAZING ASPECT OF THE BIOLOGY IS THEY HAVE NO RED BLOOD CELLS.

Is that bad?

FOR MOST VERTEBRATES IT WOULD BE **CATASTROPHIC**.

RED BLOOD CELLS CARRY THE PROTEIN **HEMOGLOBIN.** HEMOGLOBIN BINDS OXYGEN AND CARRIES IT TO THE TISSUES. ALL OTHER VERTEBRATES HAVE RED BLOOD CELLS BECAUSE WITHOUT HEMOGLOBIN, THEY WOULD DIE FROM LACK OF OXYGEN.

So how do the icefish get along without it?

THEY CAN SUR-VIVE BECAUSE THE WATER IS SO INCREDIBLY **COLD.**

COLD WATER CONTAINS MORE DISSOLVED OXYGEN THAN WARM WATER, SO THE ICEFISH CAN SOAK UP ALL THE OXYGEN THEY NEED THROUGH THEIR GILLS AND ACROSS THEIR SKIN.

O_2 O_2 O_2 O_2 O_2 O_2 O_2 O_2 O_2 O_2 O_2 O_2 O_2 O_2 O_2 O_2

MMM...

But what happened to their hemoglobin?

AT SOME POINT IN THEIR EVOLUTION, THE HEMOGLOBIN GENES IN ICEFISH MUTATED AND STOPPED WORKING. BUT, SINCE THEY DIDN'T ACTUALLY NEED THEM, THE MUTATION WASN'T LETHAL.

HEMOGLOBIN

I NO LONGER REQUIRE YOUR SERVICES. GOOD DAY.

SNIFF

OH, MAN, THAT IS COLD-BLOODED.

IN FACT, IT LOOKS LIKE THE LOSS WAS FAVORED BY NATURAL SELECTION.

Are you saying that **LOSING** a trait can be adaptive?

LOSS IS A COMMON THEME IN EVOLUTION, SIRE.

It's economics again, Dad. The loss was favored because the icefish didn't have to invest energy in making hemoglobin they didn't need.

Very well, but where is the **CONSTRAINT**, Bloort? These creatures seem ideally suited to the environment.

TO **THIS** ENVIRONMENT THEY TRULY ARE, YOUR HIGHNESS. BUT AS WE'VE SEEN, THE WORLD IS CONSTANTLY CHANGING. WHAT HAPPENS IF THE WATERS **WARM UP**?

They **ADAPT**.

Haven't you been paying attention to your own tour, Bloort?

But, Dad, the ability to adapt to changing environmental conditions requires variation in the genetic makeup of the population.

I am aware of that, Son. What's your point?

Well, if the waters warm up there will be less oxygen dissolved in them.

OXYGEN

TEMPERATURE

If that happens, there won't be enough oxygen in the water for the icefish to absorb across their skin and gills. They'll need a specialized molecule to bind the oxygen.

Ah, ah, I see. And since their genes for hemoglobin are gone, they can't get them back. They're stuck in the cold water. Well done, Son!

WELL DONE, INDEED. IF I MAY SAY SO, SIRE, I FEEL FORTUNATE TO HAVE BEEN PART OF SUCH A STUNNING, DEDUCTIVE DIALOGUE.

I ONLY WISH THAT I HAD RECORDED IT FOR THE MUSEUM SO THAT SQUINCHES EVERYWHERE AND FOR ALL TIME COULD BENEFIT FROM SUCH AN ENLIGHTENING EXCHANGE BETWEEN OUR EMPERORS OF ERUDITION.

Whew, that was a long one, Bloort.

I'M SORRY, YOUR HIGHNESS, BUT I JUST REALIZED MY FLATTERY HAS BEEN FALTERING.

Well, be careful. We don't want you to pull a muscle or something.

I'LL TRY TO PACE MYSELF, MOST GRACIOUS HIGHNESS.

THE LOSS OF THE HEMOGLOBIN GENES IS A CONSTRAINT ON THE ICEFISH. IF THE WATERS EVER WARM UP, THE ICEFISH MAY FIND THEMSELVES WITHOUT THE GENETIC TOOLS TO SURVIVE.

They will go extinct.

PERHAPS. OR PERHAPS EVOLUTION WILL SURPRISE US YET AGAIN WITH SOME NEW, UNEXPECTED INNOVATION.

You know, Bloort, as we talk about these adaptations, it makes me wonder...

SIRE?

Well, if evolution is an ongoing process, shouldn't we see some trait in **TRANSITION**? Some feature being gained or lost?

WOULD YOU LIKE TO SEE A FEW?

They exist?

OF COURSE, THERE ARE **NUMEROUS** EXAMPLES.

AS A RESULT OF THE SLOW, PIECE-MEAL TINKERING THAT IS NATURAL SELECTION, ORGANISMS OFTEN RETAIN FEATURES THAT APPEAR TO HAVE LITTLE OR NO FUNCTION. THESE TRAITS ARE CALLED **VESTIGIAL STRUCTURES.**

Are you saying that evolution leaves **LEFT-OVER BITS** behind in organisms?

IN A SENSE, YOUR MAJESTY.

That's a bit sloppy, isn't it?

ABSOLUTELY.

TAKE **SNAKES**, FOR INSTANCE. RESEARCHERS BELIEVE THAT THE LOSS OF LEGS IN THESE HIGHLY MODIFIED REPTILES MADE IT EASIER FOR THEM TO PURSUE THEIR PREY INTO BURROWS.

HE WENT THATAWAY.

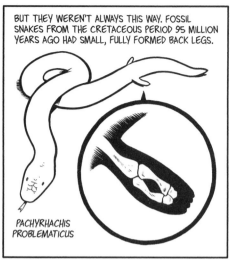

BUT THEY WEREN'T ALWAYS THIS WAY. FOSSIL SNAKES FROM THE CRETACEOUS PERIOD 95 MILLION YEARS AGO HAD SMALL, FULLY FORMED BACK LEGS.

PACHYRHACHIS PROBLEMATICUS

ALTHOUGH MOST SNAKES HAVE LONG SINCE LOST THEIR APPENDAGES, A FEW SPECIES STILL CARRY THE VESTIGES OF A BACK LIMB. THE RUBBER BOA'S SMALL BACK SPUR IS ALL THAT REMAINS OF WHAT USED TO BE LEGS.

WHAT CAN I SAY? I'M A WORK IN PROGRESS.

LIKEWISE, THE KIWI, A SMALL BIRD THAT EVOLVED TO ABANDON FLIGHT FOR A LIFE ON THE GROUND, STILL HAS LITTLE, USELESS WINGS -- A RECORD OF ITS EVOLUTIONARY PAST.

THE BLIND MOLE RAT HAS SMALL VESTIGIAL EYES. IT CAN NO LONGER USE THEM BECAUSE THEY ARE COVERED IN SKIN.

WHAT'S GOING ON?

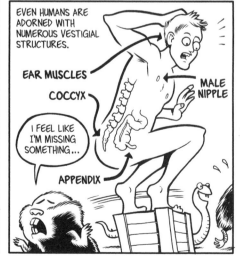

EVEN HUMANS ARE ADORNED WITH NUMEROUS VESTIGIAL STRUCTURES.

EAR MUSCLES

COCCYX

MALE NIPPLE

I FEEL LIKE I'M MISSING SOMETHING...

APPENDIX

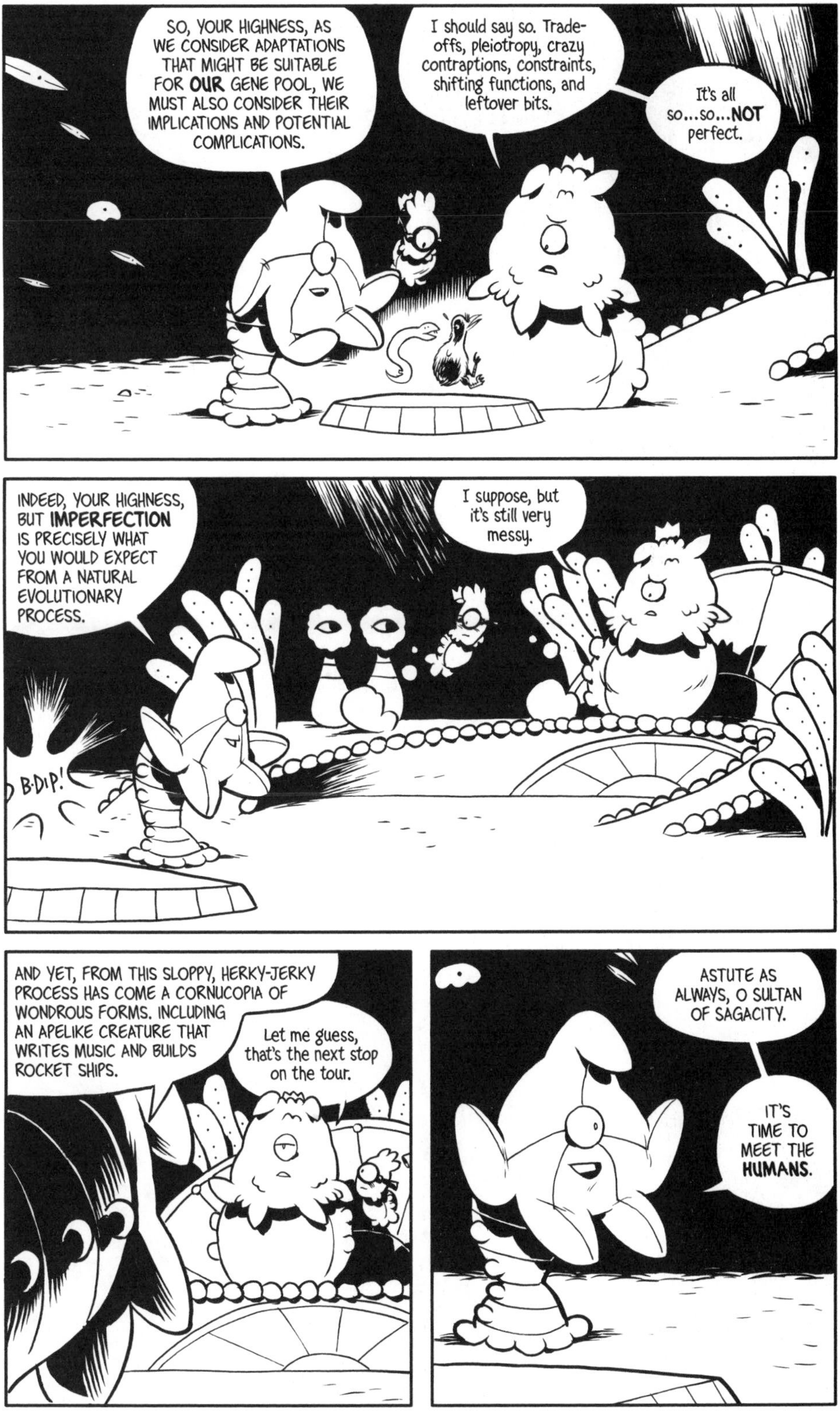
SO, YOUR HIGHNESS, AS WE CONSIDER ADAPTATIONS THAT MIGHT BE SUITABLE FOR **OUR** GENE POOL, WE MUST ALSO CONSIDER THEIR IMPLICATIONS AND POTENTIAL COMPLICATIONS.
I should say so. Trade-offs, pleiotropy, crazy contraptions, constraints, shifting functions, and leftover bits.
It's all so...so...**NOT** perfect.
INDEED, YOUR HIGHNESS, BUT **IMPERFECTION** IS PRECISELY WHAT YOU WOULD EXPECT FROM A NATURAL EVOLUTIONARY PROCESS.
I suppose, but it's still very messy.
B·DIP!
AND YET, FROM THIS SLOPPY, HERKY-JERKY PROCESS HAS COME A CORNUCOPIA OF WONDROUS FORMS. INCLUDING AN APELIKE CREATURE THAT WRITES MUSIC AND BUILDS ROCKET SHIPS.
Let me guess, that's the next stop on the tour.
ASTUTE AS ALWAYS, O SULTAN OF SAGACITY.
IT'S TIME TO MEET THE **HUMANS**.

CHAPTER 6

Getting a Leg Up on Evolution

Bloort, I thought we were going to see the humans.

WE **ARE**, YOUR MAJESTY.

Then why are we still standing in this holographic Earth water? I thought humans lived on land.

But that was hundreds of millions of years ago. I thought we were almost done with the tour.

THEY DO, YOUR HIGHNESS, BUT I WANTED TO START A LITTLE EARLIER, WHEN THE FIRST VERTEBRATES CRAWLED OUT OF THE WATER.

THIS WILL BE QUICK, MY MOST PATIENT POTENTATE. WE'RE LOOKING FOR **FISH**.

AS ALL SQUINCHES ARE INTIMATELY AWARE, THE BODIES OF AQUATIC ORGANISMS ARE SUPPORTED BY THE DENSE WATERY MEDIUM IN WHICH WE LIVE.

I THINK YOU ARE A TERRIFIC SQUINCH.

THANKS, WATER, YOU ARE **SO** SUPPORTIVE.

THE SAME IS TRUE ON EARTH. FOR FISH, THIS MEANS THEY CAN DO QUITE WELL WITH RELATIVELY SLIGHT SKELETONS AND DELICATE FINS.

BUT LIVING ON LAND REQUIRES MUCH MORE SUPPORT. LAND VERTEBRATES HAVE STURDY SKELETONS, WITH ROBUST RIB CAGES AND MEATY MUSCLES TO SUPPORT THEIR WEIGHT.

Oh, come now, Bloort, can a skeleton really be **THAT** important?

WELL, YOUR HIGHNESS, AS PART OF THE TOUR I HAVE CONSTRUCTED A TERRESTRIAL SIMULATOR TO GIVE VISITORS A SENSE OF THE DIFFERENCE BETWEEN LIVING IN WATER AND LIVING ON LAND.

Ooo, goody.

I claim royal privilege and will go first.

I HAVE FIELD GENERATORS FOR BOTH OF YOU. WE JUST NEED TO SLIP THEM AROUND YOUR WAISTS.

I'm still going first.

Okay, but I get to work the controls.

I'M NOT SO SURE THAT'S...

Ready, Dad?

Ready, Son!

BEEP.

SQUISH

Turn it off, turn it off now.

Not yet, this is hilarious.

HAHAHA

Uh-oh.

CLICK.

OH DEAR, OH DEAR, OH DEAR...

BOING

Hee heee, let's do that again!

HEH -- GULP -- A-AS YOU CAN SEE, BECAUSE WE SQUINCHES LIVE IN WATER, WE COULD THRIVE WITHOUT A BULKY INTERNAL SKELETON.

So...

...we're not well suited for living on land. Why did Earth creatures leave the ocean?

WELL, SIRE, THE LAND HAD NEW, UNEXPLOITED NICHES, AND AIR CONTAINS FORTY TIMES MORE OXYGEN THAN WATER.

So, exploration would be rewarded with access to new markets and vast oxygen wealth?

IN A MANNER OF SPEAKING, O IMPERIALISTICALLY MINDED ONE. ORGANISMS WITH ADAPTATIONS TO INVADE THE LAND WOULD HAVE SOME EXCITING EVOLUTIONARY POSSIBILITIES.

But how did they make that leap, Bloort? These fish don't seem very well suited for life in such a hostile environment.

MOST AREN'T, YOUR HIGHNESS.

THE VAST MAJORITY OF FISH ARE **RAY-FINNED** FISH AND THEIR FINS ARE LITTLE MORE THAN A FLAP OF SKIN SUPPORTED BY SOME THIN BONY SPIKES.

GASP -- I'M... ...LIKE A... GASP... FISH OUT OF WATER.

BUT DURING THE SILURIAN PERIOD -- ALMOST 450 MILLION YEARS AGO -- A SMALL GROUP CALLED THE **LOBE-FINNED** FISH PLAYED A HUGE ROLE IN THE MOVE TO LAND.

LOBE-FINNED FISH ARE CHARACTERIZED BY FLESHY FINS THAT ARE FULL OF BONES.

Nice fins, weirdo.

TODAY ONLY A FEW SPECIES SURVIVE. **COELACANTHS** LIVE IN THE WATERS OFF THE EASTERN COAST OF SOUTHERN AFRICA AND HAVE CHANGED VERY LITTLE FOR HUNDREDS OF MILLIONS OF YEARS...

EVERY MILLENNIUM IT'S THE SAME OLD THING: FLOAT, EAT, SLEEP. FLOAT, EAT, SLEEP...

...WHILE THE LUNGFISH OF AFRICA, SOUTH AMERICA, AND AUSTRALIA RETAIN MANY ANCIENT CHARACTERISTICS, SUCH AS AN ELABORATE SKELETON, BEEFY FINS TO CRAWL ACROSS LAND WHEN THEIR WATER DRIES UP, AND PRIMITIVE LUNGS TO BREATHE WATER.

THIS POND AIN'T BIG ENOUGH FOR THE BOTH OF US.

I'M A-GOIN', I'M A-GOIN'.

THE LOBE-FINNED FISH WERE NEVER AS SUCCESSFUL IN WATER AS THEIR RAY-FINNED COUSINS. THEIR EVOLUTIONARY LEGACY WOULD BE ON LAND.

Something like **THESE** fish would make the **LEAP** to land?

MORE LIKE THE **CRAWL** TO LAND, YOUR MAJESTY.

THERE WAS A THIRD, NOW EXTINCT, GROUP OF LOBE-FINNED FISH CALLED THE **TETRAPODOMORPHS** THAT WOULD BECOME THE ANCESTORS TO ALL TERRESTRIAL VERTEBRATES.

Tetra-whatta-**WHAT**?

TETRAPODOMORPH. IT MEANS "FOUR FOOT SHAPE."

THEIR TRANSITION FROM FISH TO LAND VERTEBRATE IS DETAILED EXQUISITELY IN THE FOSSIL RECORD.

BUT ONE RECENT FOSSIL DISCOVERY HAS BEEN PARTICULARLY EXCITING.

IN 2006, THE EARTH SCIENTIST NEIL SHUBIN DISCOVERED AN IMPORTANT FOSSIL CALLED **TIKTAALIK**.

Tik-TAH-lik?

Where do they come up with these names?

TIKTAALIK WAS A LOBE-FINNED FISH THAT HAD SOME KEY FEATURES THAT DISTINGUISHED IT FROM THE FISH THAT HAD PRECEDED IT. UNLIKE ANY FISH, IT HAD A **NECK**, SO IT COULD LOOK AROUND.

LOOK BOTH WAYS BEFORE CROSSING THE SHORELINE, DEAR.

AND IT HAD SPECTACULAR LEGS.

THANK YOU! I DO A HUNDRED PUSH-UPS EVERY DAY.

They really don't look that impressive to me.

HMPH!

AH, TO FULLY APPRECIATE TIKTAALIK'S LIMBS, WE MUST FIRST TAKE A LOOK AT THE LIMBS OF VARIOUS VERTEBRATES.

ALL VERTEBRATES' LIMBS SHOW THE SAME BASIC PATTERN: STARTING AT THE SHOULDER, WE SEE -- AS PROFESSOR SHUBIN POINTS OUT -- ONE BONE, TWO BONES, A BLOB OF LITTLE BONES, AND THEN FIVE DIGITS.

THIS IS TRUE FOR HUMANS...

HUMERUS

ULNA

RADIUS

HUMAN

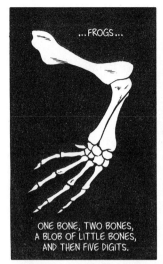

...FROGS...

ONE BONE, TWO BONES, A BLOB OF LITTLE BONES, AND THEN FIVE DIGITS.

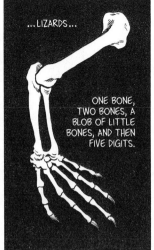

...LIZARDS...

ONE BONE, TWO BONES, A BLOB OF LITTLE BONES, AND THEN FIVE DIGITS.

Interesting, but...

WE SEE IT REPEATED AGAIN AND AGAIN AND AGAIN!

CAT

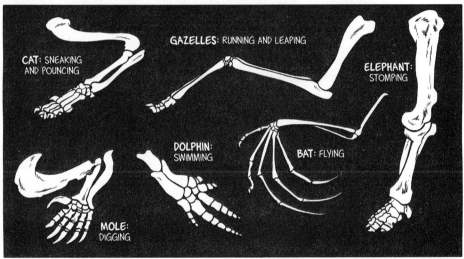

CAT: SNEAKING AND POUNCING

GAZELLES: RUNNING AND LEAPING

ELEPHANT: STOMPING

MOLE: DIGGING

DOLPHIN: SWIMMING

BAT: FLYING

HUMAN

These limbs have very different functions, but they are constructed using the same basic plan.

THAT'S RIGHT.

THE SIMILARITIES YOU SEE IN THE LIMBS OF VERTEBRATES ARE CALLED HOMOLOGIES.

A **HOMOLOGOUS** TRAIT IS A TRAIT THAT TWO OR MORE SPECIES INHERITED FROM A COMMON ANCESTOR. HOMOLOGIES CAN BE ANATOMICAL, GENETIC, OR DEVELOPMENTAL.

ONE OF THE REASONS TIKTAALIK IS IMPORTANT...

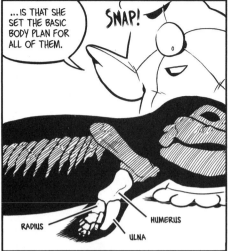

...IS THAT SHE SET THE BASIC BODY PLAN FOR ALL OF THEM.

SNAP!

RADIUS

HUMERUS

ULNA

So you're saying that all vertebrates have the same pattern of bones in their limbs because they all descended from a common ancestor who had that pattern?

I HAVE TO "HAND" IT TO YOU, YOUR HIGHNESS, YOU ARE REALLY GETTING A "LEG UP" ON THIS MATERIAL.

That was AWFUL, Bloort.

giggle

Hmm, I prefer it when he grovels.

AS YOU WISH, SIRE. YOUR PATIENCE WITH MY FEEBLE WITTICISMS IS EXCEEDED ONLY BY THE VASTNESS OF YOUR INFINITE INTELLECTUAL ACUMEN.

That's more like it.

TIKTAALIK HAD AN ADAPTATION THAT DRAMATICALLY INCREASED HER EVOLUTIONARY SUCCESS, AND SHE PASSED IT ALONG TO ALL OF HER DESCENDANTS.

THE VERTEBRATE LIMB HAS BEEN AN ENORMOUSLY IMPORTANT AND VERSATILE ADAPTATION. IT HAS EVOLVED INTO A DAZZLING ARRAY OF VARIATIONS, BUT EACH IS STILL JUST A VARIATION ON THE SAME THEME.

THE HIGH DEGREE OF HOMOLOGY BETWEEN THE BONES OF VERTEBRATES AND HUMANS SPARKED MANY EARTH SCIENTISTS TO SPECULATE ON HOW THEY THEM-SELVES WERE RELATED TO OTHER ANIMALS.

TONIGHT
◇ o ◇
ARE
HUMANS
RELATED TO
APES
◇ o ◇
T. H. Huxley

DARWIN'S CONTEMPORARY **T. H. HUXLEY** BECAME FAMOUS FOR GIVING PUBLIC LECTURES ABOUT THE SIMILARITIES BETWEEN HUMAN SKELETONS AND THE SKELETONS OF THEIR NEAREST PRIMATE RELATIVES.

IT IS QUITE CERTAIN THAT THE APE WHICH MOST NEARLY APPROACHES MAN IS EITHER THE CHIMPANZEE OR THE GORILLA...

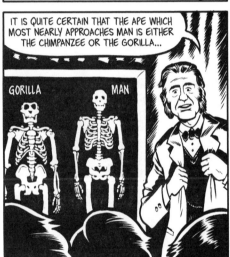

GORILLA MAN

Whoa, whoa, whoa, Bloort. You're getting ahead of us with the terminology. Primate? Chimpanzee? Gorilla?

APOLOGIES, YOUR HIGHNESS. PERHAPS IT WOULD BE BEST TO TAKE A STEP BACK AND CON-SIDER WHERE HUMANS SIT IN RELATION TO OTHER ANIMALS.

FIRST, WE KNOW THAT HUMANS ARE **MULTICELLED EUKARYOTES.** ALL EUKARYOTES SHARE SOME FUNDA-MENTAL FEATURES: THEY HAVE CELLS WITH NUCLEI AND LINEAR DNA.

EUKARYOTES

119

HUMANS ARE **ANIMALS**, WHICH MEANS THEY MOVE AROUND AND EAT OTHER THINGS. THEY CANNOT MAKE THEIR OWN FOOD LIKE PLANTS CAN.

ANIMALS

WITHIN THE ANIMALS, HUMANS ARE IN A GROUP CALLED **MAMMALS**.

MAMMALS

The furry little creatures with mammary glands and biggish brains?

EXCELLENT MEMORY, SIRE. MAMMALS CAN BE FURTHER SORTED INTO EVEN SMALLER, SPECIALIZED CATEGORIES BASED UPON THEIR HOMOLOGOUS TRAITS.

HUMANS ARE PART OF A SUBSET OF MAMMALS CALLED **PRIMATES**. THE PRIMATES ARE PRIMARILY TREE-DWELLING MAMMALS THAT INCLUDE MONKEYS, LEMURS, TARSIERS, AND APES.

PRIMATES

WITHIN THE PRIMATES, HUMANS ARE PART OF AN EVEN SMALLER GROUP OF LARGE, TAILLESS PRIMATES CALLED **APES**.

APES

Hmph. They all look the same to me.

AH, BUT I'M SURE THAT YOUR KEEN ROYAL EYES CAN FIND DIFFERENCES IF YOU LOOK HARD ENOUGH, SIRE.

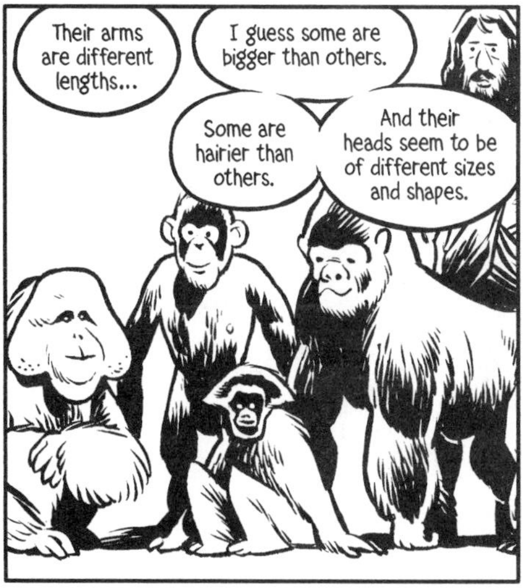
Their arms are different lengths...
I guess some are bigger than others.
Some are hairier than others.
And their heads seem to be of different sizes and shapes.

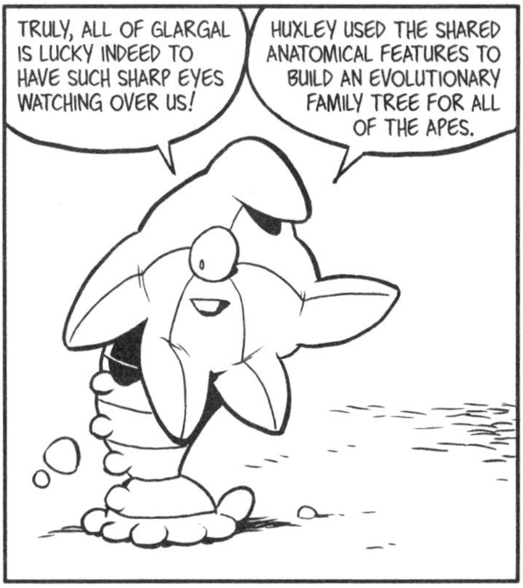
TRULY, ALL OF GLARGAL IS LUCKY INDEED TO HAVE SUCH SHARP EYES WATCHING OVER US!
HUXLEY USED THE SHARED ANATOMICAL FEATURES TO BUILD AN EVOLUTIONARY FAMILY TREE FOR ALL OF THE APES.

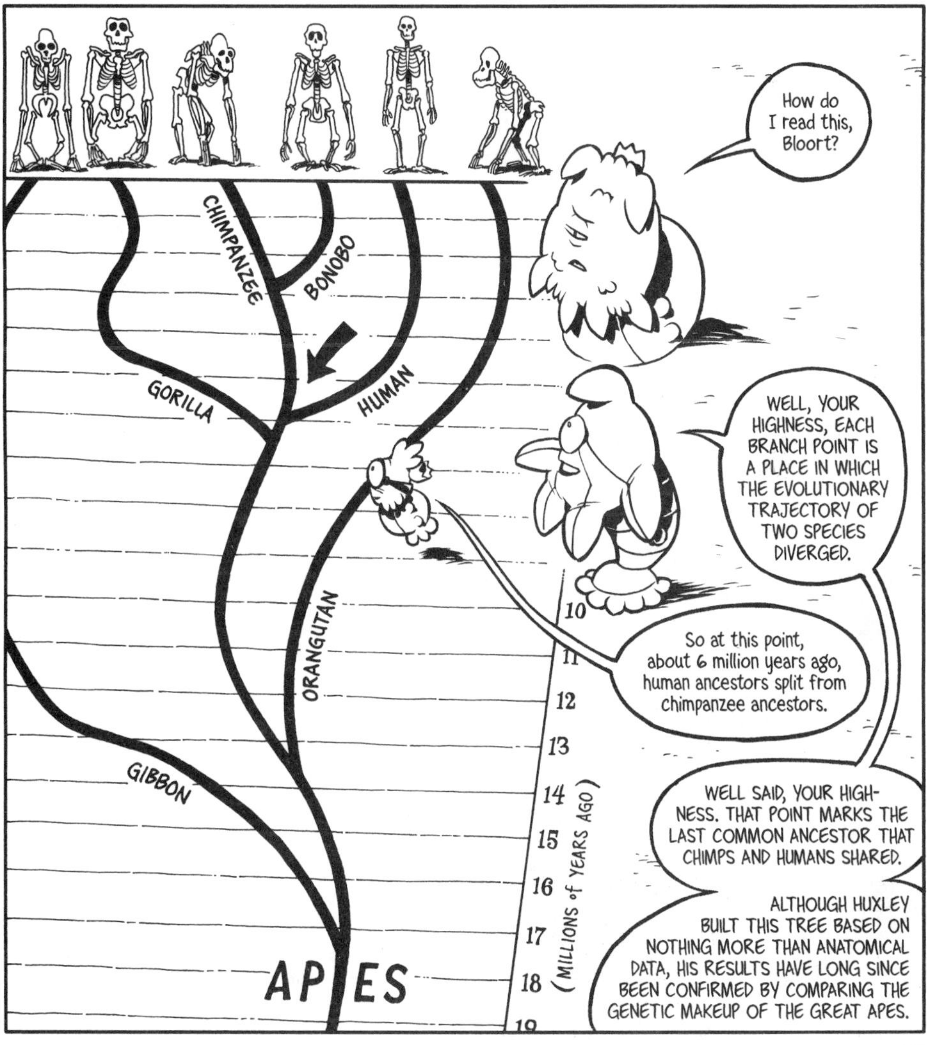
How do I read this, Bloort?
WELL, YOUR HIGHNESS, EACH BRANCH POINT IS A PLACE IN WHICH THE EVOLUTIONARY TRAJECTORY OF TWO SPECIES DIVERGED.
So at this point, about 6 million years ago, human ancestors split from chimpanzee ancestors.
WELL SAID, YOUR HIGHNESS. THAT POINT MARKS THE LAST COMMON ANCESTOR THAT CHIMPS AND HUMANS SHARED.
ALTHOUGH HUXLEY BUILT THIS TREE BASED ON NOTHING MORE THAN ANATOMICAL DATA, HIS RESULTS HAVE LONG SINCE BEEN CONFIRMED BY COMPARING THE GENETIC MAKEUP OF THE GREAT APES.
CHIMPANZEE
BONOBO
GORILLA
HUMAN
ORANGUTAN
GIBBON
APES
10
11
12
13
14
15
16
17
18
(MILLIONS of YEARS AGO)

CHIMPS AND HUMANS SHARE 98% HOMOLOGY OF THEIR DNA, WHILE HUMANS AND GORILLAS SHARE 97% HOMOLOGY.

That's not much difference.

OH, BUT YOUR HIGHNESS, HAVE WE NOT SEEN DRAMATIC ANATOMICAL DIFFERENCE ARISE FROM SMALL GENETIC CHANGES?

Remember those male isopods that had three phenotypes due to variation in **ONE** gene, Dad?

Ah, yes, and Belyaev's doggy-foxes were the product of selection on a **SINGLE** gene.

BONOBO

CHIMPANZEE

HUMAN

6 MILLION YEARS AGO

COMMON ANCESTOR

THE 98% GENETIC HOMOLOGY TELLS US THAT SPECIES ARE CLOSELY RELATED. BUT A 2% DIFFERENCE CAN LEAD TO SOME **SIGNIFICANT** VARIATION.

WITHIN THAT VARIATION ARE THE ADAPTATIONS THAT ALLOWED HUMANS TO BECOME... WELL...

... HUMAN.

AS WE'VE SAID, THE HUMAN EVOLUTIONARY LINE AND THE CHIMP EVOLUTIONARY LINE SPLIT ABOUT 6 MILLION YEARS AGO. BUT HUMANS WEREN'T HUMANS 6 MILLION YEARS AGO. THESE ANCESTORS TO MODERN HUMANS WERE CALLED **HOMINIDS**.

OOO OOOK!*

OO EE OO!**

* FARE THEE WELL.

** ALAS, PARTING IS SUCH SWEET SORROW.

ABOUT 1.6 MILLION YEARS AFTER THE HOMINDS SPLIT FROM THE CHIMPS, THEY DID SOMETHING **REVOLUTIONARY**.

Wait, wait, don't tell me. They invented jet packs.

UH, NO, SIRE, THEY...

Think, Son. They didn't have the technology to make rocket fuel.

Bloort probably means they built some kind of wheeled device that they propelled with feet.

WHAT? NO... UM --

122

ACTUALLY, THE REVO- LUTIONARY THING THEY DID WAS... **STAND UP.**

And then strap on a jet pack?

NO, JUST STAND UP.

Bloort, we need to discuss your definition of "revolutionary."

BUT IT **WAS**, YOUR HIGHNESS. IN SO MANY UNIMAGINABLE WAYS!

VERY RECENTLY, EARTH SCIENTIST TIM WHITE AND A TEAM OF EXPERTS DISCOVERED ONE OF THE OLDEST HOMINID FOSSILS EVER FOUND.

ARDIPITHECUS RAMIDUS -- NICKNAMED **ARDI** -- LIVED 4.4 MILLION YEARS AGO.

So that would have been about 1.6 million years **AFTER** the hominid and chimp lines split.

CORRECT, O NUMERICALLY NIMBLE ONE. AND IN THAT TIME, ARDI EVOLVED SOME INTERESTING TRANSITIONAL FEATURES.

FIRST, HER FOOT BECAME STIFFER THAN A CHIMP'S. A RIGID FOOT ACTS AS A LEVER THAT MAKES WALKING ON TWO FEET EASIER. MODERN HUMANS ALSO HAVE A RIGID FOOT.

MY FOOT'S KIND OF STIFF.

EXCELLENT. LET'S GO FOR A WALK.

BUT IN SOME WAYS, ARDI'S FOOT IS A SNAPSHOT OF A FOOT IN EVOLUTIONARY TRANSITION BECAUSE IT RETAINED AN **OPPOSABLE BIG TOE.** THIS PROBABLY MADE IT EASIER FOR HER SPECIES TO CLIMB THE TREES OF HER FOREST HOME.

I MEANT A WALK ON THE **GROUND.**

WALKING ON TWO LEGS SET IN MOTION A SERIES OF EVOLUTIONARY DOMINOES FOR THE HOMINIDS.

KICK!

MODIFIED TOES
MODIFIED LEGS
MODIFIED PELVIS
MODIFIED BACKBONE
MODIFIED SKULL

LESS THAN A MILLION YEARS AFTER ARDI, ANOTHER HOMINID CALLED *AUSTRALOPITHECUS AFARENSIS*-- NICKNAMED **LUCY** -- WOULD EVOLVE A FOOT VIRT- UALLY INDISTINGUISHABLE FROM MODERN HUMANS'.

THESE FEET ARE MADE FOR **WALKIN'**.

HOW ARE THEY FOR **RUNNING**?

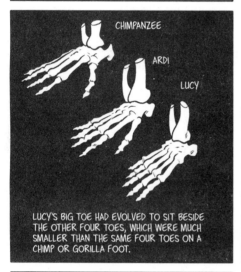

CHIMPANZEE

ARDI

LUCY

LUCY'S BIG TOE HAD EVOLVED TO SIT BESIDE THE OTHER FOUR TOES, WHICH WERE MUCH SMALLER THAN THE SAME FOUR TOES ON A CHIMP OR GORILLA FOOT.

LUCY'S SPECIES ALSO EVOLVED AN ARCH AND A LONG HEEL TO ABSORB THE SHOCK OF WALKING. THESE TWO CHANGES ALLOWED THEM TO WALK FOR VERY LONG TIMES COMPARED TO OTHER APES.

ARE WE THERE YET?

SIGH

WALKING UPRIGHT FOR LONG PERIODS OF TIME MEANT OUR **PELVISES** NEEDED TO BECOME MORE **BOWL-SHAPED** TO SUPPORT OUR GUTS.

SHH. WE HAVEN'T EVOLVED LANGUAGE YET.

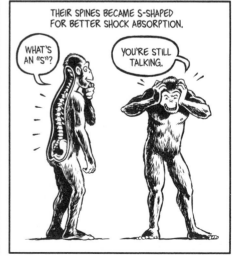

THEIR SPINES BECAME S-SHAPED FOR BETTER SHOCK ABSORPTION.

WHAT'S AN "S"?

YOU'RE STILL TALKING.

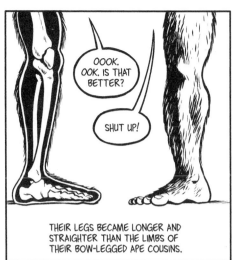

OOOK. OOK. IS THAT BETTER?

SHUT UP!

THEIR LEGS BECAME LONGER AND STRAIGHTER THAN THE LIMBS OF THEIR BOW-LEGGED APE COUSINS.

AND THEIR SKULLS NOW SAT ON TOP OF THEIR SPINES.

IT'S VERY NICE UP THERE. LOTS OF ROOM TO GROW. CAN LANGUAGE BE FAR BEHIND?

NO, BUT UNTIL THEN, NO TALKING!

THE RESULT WAS AN ORGANISM VERY WELL SUITED FOR LIVING LIFE ON TWO LEGS.

But wasn't all that walking **EX-HAUSTING**?

ACTUALLY, YOUR HIGHNESS, SCIENTISTS HAVE COMPARED THE ENERGY REQUIRED TO WALK ON TWO LEGS WITH THE ENERGY REQUIRED TO WALK ON ALL FOURS LIKE A CHIMP. WALKING ON TWO LEGS USES **75% LESS ENERGY!**

My word, such an energy savings would be a tremendous advantage!

INDEED, YOUR HIGHNESS. BUT IT ALSO DID SOMETHING MORE SIGNIFICANT. IT FREED THEIR **HANDS** TO EXPLORE THEIR WORLD.

HOMINIDS USED THEIR HANDS TO POKE AND PROD THE WORLD LIKE NO OTHER CREATURE BEFORE THEM. THEIR HANDS EVOLVED AGILE OPPOSABLE THUMBS THAT ALLOWED THEM TO GRASP, CARRY, AND SHAPE OBJECTS.

KNOCK IT OFF, OR I'M TELLIN' MOM.

OKE POKE POKE POKE POKE

But...this chimp's thumb looks opposable, too, Bloort.

PICK!

IT IS, SIRE. BUT THE CHIMP'S THUMB COULD TOUCH ONLY ITS FIRST TWO FINGERS. THE HOMINID THUMB COULD TOUCH WHAT HUMANS CALL THEIR RING FINGER AND THE PINKY.

WINK

ALONG WITH THE THUMB CAME ADAPTATIONS IN THE TENDONS OF THE HAND THAT ALLOWED THE WRIST TO SWIVEL AND MOVE FAR MORE THAN IN APES.

oooOo AAH!

THIS ADDED DEXTERITY WOULD BE A TREMENDOUS ADVANTAGE BECAUSE IT GAVE SOME HOMINID SPECIES THE ABILITY TO MAKE **TOOLS**.

Surely **OTHER** creatures on Earth use tools, Bloort.

OH, INDEED THEY **DO**, YOUR HIGHNESS. MANY SPECIES USE THEM. CHIMPS USE LEAVES OR GRASS TO PULL TERMITES FROM THEIR MOUNDS...

FINALLY. IT IS SO NICE TO GET OUT OF THAT STUFFY COLONY FOR A WHILE.

...EGYPTIAN VULTURES USE ROCKS TO CRACK OPEN OSTRICH EGGS...

HOW WOULD YOU LIKE YOUR EGGS, DEAR?

CRACKED OPEN AND RUNNY, PLEASE.

...AND WHEN DOLPHINS FORAGE ON THE SEA FLOOR, SOME HOLD SPONGES IN THEIR MOUTHS TO PROTECT THEIR NOSES FROM GETTING ALL SCRATCHED UP.

IT MAY LOOK WEIRD, BUT IT'S FOR A GOOD PORPOISE.

BUT, IN THESE AND OTHER CASES, THE ANIMALS ARE USING OBJECTS THEY'VE **FOUND** TO PERFORM A TASK.

ONLY THE HOMINIDS EVOLVED SPECIES LIKE *HOMO HABILIS*, WHICH MEANS "HANDY MAN," THAT COULD TRANSFORM AN OBJECT LIKE A **STONE**...

...INTO A **KNIFE**.

AND THEN ADD A STICK TO MAKE IT A **SPEAR**.

HOMO HABILIS

THESE TOOLS WOULD BECOME ESSENTIAL TO THEIR SUCCESS IN GETTING MEAT TO EAT. AT FIRST, THEY PROBABLY JUST ATE WHAT THEY COULD SCAVENGE...

SIGH --

LEFTOVERS AGAIN?

...BUT EVENTUALLY THEY STARTED USING TOOLS TO HUNT THEIR FOOD WITH.

THIS FAST FOOD IS KILLING ME.

THE HIGH PROTEIN AND FAT CONTENT OF MEAT MAY HAVE FUELED SIGNIFICANT GROWTH IN THE SIZE OF SOME HOMINID BRAINS. AND WHEN THEY INVENTED COOKING THEY COULD EAT EVEN MORE BECAUSE THE MEAT BECAME EASIER TO DIGEST!

Y'KNOW WHAT WE NEED TO DO? WE NEED TO INVENT S'MORES.

AS THESE WELL-FED HOMINIDS USED THEIR NIMBLE THUMBS AND HANDS TO MAKE TOOLS, THEIR GROWING MINDS WERE ALSO BEING RESHAPED.

?

EARTH SCIENTISTS ARE NOW DIS-COVERING THAT THE PART OF THE BRAIN THAT CONTROLS LANGUAGE DEVELOPED FROM THE PART OF THE BRAIN THAT CONTROLS HUMAN HANDS.

If I understand what you're saying, Bloort, limbs have had a huge effect on human evolution.

Adaptations in hominid feet and legs allowed them to stand up.

TA-DAAA!

127

This freed their hands from walking and led to the evolution of highly opposable thumbs and terrific dexterity.

SOME PUT THAT DEXTERITY TO BETTER USE THAN OTHERS.

WHAT?

This made it possible for some hominids to make tools, which in turn helped reshape their brains and made language possible.

WELL SAID, AS ALWAYS, YOUR HIGHNESS. EVENTUALLY, ADAPTATIONS IN THE HUMAN **PHARYNX** ALLOWED THEM TO MAKE SOPHISTICATED VOCALIZATIONS TO COMMUNICATE THEIR IDEAS.

Don't the other apes have a pharynx, Bloort?

THEY DO, SIRE. BUT THE HUMAN PHARYNX IS POSITIONED LOWER IN THE THROAT AND, UNLIKE IN THEIR CHIMP COUSINS, THE HUMAN PHARYNX AND BREATHING TUBE IS RIGHT BESIDE THE TUBE HUMANS USE TO EAT. CONSEQUENTLY, HUMANS CANNOT BREATHE AND SWALLOW AT THE SAME TIME.

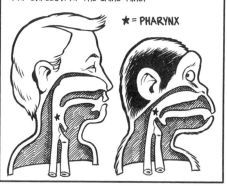

★ = PHARYNX

BEFORE THE INVENTION OF A PROCEDURE CALLED THE **HEIMLICH MANEUVER**, CHOKING WAS THE 6TH LEADING CAUSE OF DEATH IN HUMANS.

ARE YOU OKAY?

HURCK!!

I'M FINE.

NOW, WHAT WAS I SAYING?

It is a trade-off! They get to speak, but it comes with a price.

I STAND IN AWE OF YOUR INEXHAUSTIBLE INTELLECT, YOUNG SIRE. TRULY YOU HAVE AN ASTONISHING ABILITY TO DRAW TOGETHER THE MANY THREADS OF THIS TOUR.

So they now can walk, talk, create, and share ideas. What an amazing end to their evolutionary journey.

BUT THAT ISN'T THE **END**, SIRE.

Yes, **YES**, of course, they got more sophisticated with their toolmaking and eventually visited their **MOON**.

What I mean is, now that they're intelligent, self-aware beings, they've stopped evolving.

BUT THAT'S JUST IT. THEY **HAVEN'T** STOPPED EVOLVING, SIRE. THE ONLY SPECIES THAT STOP EVOLVING ARE, WELL...

...extinct!

AND HUMANS ARE FAR FROM THAT.

WITH THE ADVENT OF AGRICULTURE 10,000 YEARS AGO, HUMANS BEGAN TO ADAPT TO NEW FOODS.

FOR EXAMPLE, MOST MAMMALS CONSUME MILK ONLY AS INFANTS. AS THEY GROW OLDER, THEY CAN NO LONGER DIGEST A MILK SUGAR CALLED **LACTOSE**.

ICK!

I DON'T KNOW HOW THEY DRINK THAT STUFF.

BUT HUMAN POPULATIONS THAT DEPEND ON LIVE-STOCK FOR THEIR SURVIVAL HAVE EVOLVED GENES THAT ALLOW THEM TO DIGEST LACTOSE AS ADULTS.

GOOD THING, TOO. MY LATTE WOULDN'T BE MUCH WITHOUT MY FOAMED MILK.

I LIKE VANILLA.

So, it was an advantage to be able to digest that sugar?

ABSOLUTELY, YOUR HIGH-NESS. ALL EARTH ORGANISMS RELY ON SUGARS TO FUEL THEIR BODIES.

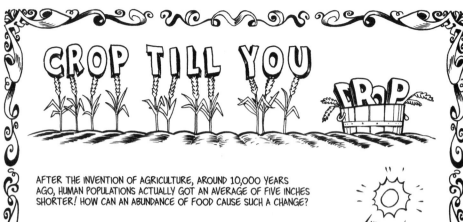

CROP TILL YOU CROP

AFTER THE INVENTION OF AGRICULTURE, AROUND 10,000 YEARS AGO, HUMAN POPULATIONS ACTUALLY GOT AN AVERAGE OF FIVE INCHES SHORTER! HOW CAN AN ABUNDANCE OF FOOD CAUSE SUCH A CHANGE?

EARLY HUMANS WERE ADAPTED TO LIFE AS HUNTER-GATHERERS AND THEY ATE A LOT OF MEAT RICH IN VITAMINS AND NUTRIENTS. WHEN HUMANS SHIFTED TO AN AGRICULTURAL LIFE, THEIR MEAT CONSUMPTION DRASTICALLY DECREASED AND THEY FOUND THEMSELVES IN AN ENVIRONMENT TO WHICH THEY WERE NOT WELL SUITED.

FARMING CREATED A LOT OF FOOD, BUT THAT FOOD HAD FEWER NUTRIENTS THAN A MOSTLY MEATY DIET. AS A RESULT, THE NEW DIET DROVE THE EVOLUTION OF SMALLER CRANIUMS, SHORTER BODIES, AND LIGHTER BONES. IN NORTHERN LATITUDES, IT MAY ALSO HAVE CONTRIBUTED TO A DIVERSITY OF SKIN COLORS.

ONE OF THE NUTRIENTS FOUND IN MEAT THAT'S MISSING IN THESE NEW CROPS WAS VITAMIN D, WHICH IS NEEDED FOR KEEPING BONES STRONG AND HEALTHY. BUT MEAT IS NOT THE ONLY SOURCE OF VITAMIN D. IT CAN ALSO BE MADE IN A PERSON'S SKIN BY ABSORBING ULTRAVIOLET RAYS FROM THE SUN. LIGHTER SKIN MAKES IT POSSIBLE TO ABSORB MORE SUNLIGHT. PEOPLE LIVING IN THE NORTH GET LESS SUN DURING THE YEAR THAN PEOPLE LIVING IN THE SOUTH, SO NATURAL SELECTION FAVORED LIGHTER SKIN AMONG THE NORTHERN FARMERS.

HUMAN POPULATIONS ALSO EVOLVED NEW GENES TO MAKE THE ENZYMES NEEDED TO DIGEST THE ABUNDANT CARBOHYDRATES IN THE GRAINS THEY WERE GROWING. WHEN CARBOHYDRATES ARE BROKEN DOWN, THEY PRODUCE SUGARS. NEW GENES EVOLVED FOR REGULATING THE HORMONE INSULIN, WHICH STIMULATES CELLS IN THE BODY TO ABSORB THOSE SUGARS.

SOME RESEARCHERS BELIEVE THAT AGRICULTURE ACCELERATED HUMAN EVOLUTION BY ONE HUNDRED TIMES IN THE LAST 10,000 YEARS. IF THEY ARE RIGHT, THEN WE TRULY ARE WHAT WE EAT.

WHEN COMPARING HUMANS TO THEIR CLOSEST RELATIVES, THE CHIMPS, WE CAN SEE OTHER RECENT DRAMATIC CHANGES. CONSIDER THE PROTEIN ENZYME **AMYLASE**, WHICH BREAKS DOWN STARCH INTO GLUCOSE, THE PRIMARY SUGAR THE BODY CAN USE.

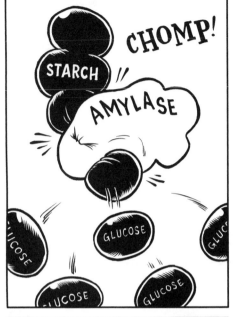

CHIMPS HAVE ONE GENE TO MAKE AMYLASE, BUT HUMANS CAN HAVE UP TO TEN COPIES. THESE EXTREME CASES ARE SEEN IN POPULATIONS THAT RELY HEAVILY ON STARCHY FOODS -- SUCH AS RICE -- FOR THEIR SURVIVAL.

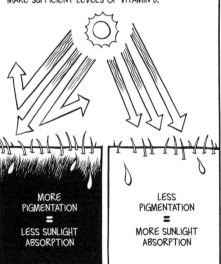

All this talk of food is making me hungry, Bloort. Are there any nondietary examples?

THERE ARE A NUMBER OF EXAMPLES, YOUR HIGHNESS.

AS THE HUMANS MOVED OUT OF AFRICA INTO EUROPE, IT BECAME ADAPTIVE TO LOSE MUCH OF THEIR PIGMENTATION.

Yikes. They look kind of pasty. Is that healthy?

ACTUALLY, SIRE, IT WAS AN IMPORTANT ADAPTATION. HUMAN SKIN ABSORBS ULTRAVIOLET RADIATION FROM SUNLIGHT AND USES IT TO MAKE **VITAMIN D**, A MOLECULE CRUCIAL TO THEIR GOOD HEALTH.

AS HUMANS MOVED NORTH INTO EUROPE, THEY EXPERIENCED LESS DIRECT SUNLIGHT. THIS DROVE THE LOSS OF PIGMENTATION IN THESE POPULATIONS BECAUSE PALE SKIN CAN ABSORB MORE LIGHT TO MAKE SUFFICIENT LEVELS OF VITAMIN D.

MORE PIGMENTATION = LESS SUNLIGHT ABSORPTION

LESS PIGMENTATION = MORE SUNLIGHT ABSORPTION

HUMAN POPULATIONS IN TIBET LIVE HIGH IN THE HIMALAYA MOUNTAINS, WHERE OXYGEN IS THINNER THAN IT IS AT LOWER ALTITUDES.

IN RESPONSE TO THIS ENVIRONMENTAL CHALLENGE, A MUTATED FORM OF THE MOLECULE HEMOGLOBIN HAS EVOLVED IN TIBETAN POPULATIONS. TIBETAN HEMOGLOBIN BINDS OXYGEN MORE TIGHTLY THAN THE HEMOGLOBIN FOUND AT LOWER ALTITUDES.

O_2

HEMOGLOBIN

IT'S MINE, ALL MINE!!

O_2

O_2

O_2

O_2

Hemoglobin? That oxygen-grabbing molecule that the icefish have lost?

THE SAME, YOUR HIGHNESS.

My goodness, these Earth creatures really **ARE** all related, aren't they?

And **STILL EVOLVING.**

YES, YOUR HIGHNESS.

BY SOME ESTIMATES, HUMAN EVOLUTION IS PROCEEDING 100 TIMES FASTER THAN IT DID BEFORE THE ADVENT OF AGRICULTURE.

BIP

BOOP

Astonishing.

There is so much to take in, Bloort.

I agree, but the adaptability of life on Earth... the dramatic power of small genetic changes...the humans' ability to manipulate the evolution of another species...it gives me hope that we can find a solution to our genetic crisis, Bloort.

Absolutely. If humans can manipulate the evolutionary trajectory of other species, surely we squinches can do the same for ourselves.

And it wouldn't necessarily require a huge genetic change, perhaps just the right gene or two.

We could explore the diversity of squinches and see if the answer isn't somewhere in our populace.

And if it isn't, perhaps we can **ENGINEER** the changes we need to avoid extinction.

SMACK!

This holographic museum is a true service to the squinches of Glargal. You should give Bloort a big grant, Dad.

YOUR PRAISE OVERWHELMS MY HUMBLE, GRASPING MIND, O MOST MAGNANIMOUS ONE.

Right, well, we better get going, then. Lots to think about, lots to think about...

RIGHT THIS WAY, SIRE. WE HAVE BUT ONE STOP LEFT ON OUR TOUR.

Another one? Seriously, Bloort, I'm getting hungry.

SMALL SNACK CAKES ARE PROVIDED, MOST ROBUST ONE.

Well, in **THAT** case, lead on.

EPILOGUE

The Path Not Taken

These cakes are very nice, Bloort.

Uh, Bloort, who's the new guy??

AH, THAT IS AN EARTH SEA CUCUMBER.

But it's just standing...

FLUMP.

Er...**LYING** there.

Have a little dignity, sir. Honestly, Bloort, what went wrong with these Earth sea cucumbers?

WHAT DO YOU MEAN, SIRE?

SEA CUCUMBERS HAVE BEEN AROUND FOR OVER 400 MILLION YEARS.

THEY ARE A GREAT SUCCESS. THEY JUST NEVER EVOLVED INTELLIGENCE. OR AN EYE. OR THE ABILITY TO SPEAK. OR THE CAPACITY TO LIFT AND MANIPULATE OBJECTS. OR --

But how could they look so similar to us yet have taken such a different path?

THE SIMILARITIES YOU SEE ARE DUE TO **CONVERGENT EVOLUTION,** YOUR HIGHNESS. CONVERGENT EVOLUTION OCCURS WHEN TWO UNRELATED SPECIES EVOLVE SIMILAR SOLUTIONS TO SUCCEED IN A SIMILAR ENVIRONMENT.

FOR EXAMPLE, **ICHTHYOSAURS** EVOLVED FROM REPTILES, AND **DOLPHINS** EVOLVED FROM MAMMALS. BUT THEY BOTH HAVE STREAMLINED BODIES, WITH DORSAL FINS AND FLIPPERS.

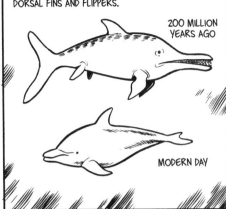

200 MILLION YEARS AGO

MODERN DAY

Does that mean their flippers are **NOT** homologous traits?

THAT'S RIGHT, SIRE. ICHTHYOSAURS AND DOLPHINS DO NOT SHARE A COMMON ANCESTOR THAT HAD FLIPPERS. THEY EACH EVOLVED FROM LAND ANIMALS WITH LEGS.

Well, if I'm following your reasoning, then the similarity in appearance between squinch and sea cucumbers may be because we both live in an aquatic environment.

PRECISELY, SIRE. WHEN WE PEER BENEATH THE SUPERFICIAL SIMILARITIES, WE SEE SOME STARTLING DIFFERENCES.

DESPITE HAVING A FRONT END, SEA CUCUMBERS NEVER EVOLVED **BRAINS** IN THE FRONT, AS WE HAVE.

Oh, I just thought it was shy.

THEY BREATHE THROUGH THEIR BOTTOMS.

That explains the bad breath.

AND WHEN THEY ARE DISTURBED, THEY WILL EJECT THEIR INTERNAL ORGANS TO DISSUADE POTENTIAL PREDATORS.

BLORK!

How **RUDE**.

Doesn't that **KILL** them, Bloort?

NO, THEY EVENTUALLY REGROW EVERYTHING.

Do they at least like sweet snack cakes?

I'M AFRAID NOT, SIRE. THEY EXTRACT SMALL BITS OF ORGANIC MATERIAL FROM **MUD**.

POKE

So, you're saying there is nothing **INHERENTLY** superior about our form? There's nothing about our body type that leads unavoidably to our grand squinchness?

I'M AFRAID NOT, SIRE.

BECAUSE THE **EVOLUTION** OF ALL LIFE FORMS DEPENDS ON **RANDOM MUTATIONS**, THE GREAT EARTH EVOLUTIONARY BIOLOGIST STEPHEN JAY GOULD ONCE POSTULATED THAT IF WE COULD REWIND THE TAPE OF LIFE AND START OVER, LIFE WOULD EVOLVE VERY DIFFERENTLY.

GOULD

That's an intriguing idea, Bloort, but how do you test it? I've learned enough at this point to expect some **DATA** to support your claims.

YES -- THERE **IS** DATA, YOUR HIGHNESS.

Nuts.

I was hoping we were almost done...

TWENTY YEARS AGO RICHARD LENSKI STARTED AN EVOLUTIONARY EXPERIMENT WITH A SINGLE *E. COLI* BACTERIUM.

READY WHEN YOU ARE, DOC.

LENSKI

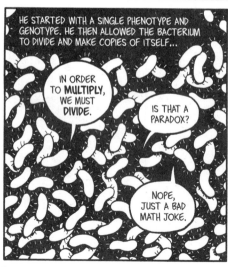

HE STARTED WITH A SINGLE PHENOTYPE AND GENOTYPE. HE THEN ALLOWED THE BACTERIUM TO DIVIDE AND MAKE COPIES OF ITSELF...

IN ORDER TO **MULTIPLY**, WE MUST **DIVIDE**.

IS THAT A PARADOX?

NOPE, JUST A BAD MATH JOKE.

...AND THEN USED THESE IDENTICAL COPIES TO START TWELVE ISOLATED POPULATIONS OF *E. COLI*.

No gene flow between any of the populations?

NONE, YOUR HIGHNESS.

POP. 10

POP. 11

POP. 12

HE FED THE E. COLI A TINY BIT OF THE SUGAR **GLUCOSE** EVERY DAY AND ALLOWED THEM TO GROW, GENERATION AFTER GENERATION. UNDER THESE CONDITIONS, THE BACTERIA POPULATIONS WENT THROUGH SIX AND A HALF GENERATIONS PER DAY.

GLUCOSE

EVERY 500 GENERATIONS, HE WOULD TAKE A FEW BACTERIA FROM EACH OF THE TWELVE POPULATIONS AND FREEZE THEM.

FREEZER

NOT FOR FOOD!

Those samples would kinda be like genetic fossils of each population, huh, Bloort?

EXACTLY! TO DATE, DR. LENSKI'S BACTERIA HAVE LIVED MORE THAN **44,000 GENERATIONS** AND HE CAN USE THE GENETIC FOSSILS COLLECTED EVERY 500 GENERATIONS TO MONITOR THE E. COLI'S EVOLUTION.

OVER THE LAST TWENTY YEARS, THE TWELVE POPULATIONS OF E. COLI DESCENDED FROM THAT ORIGINAL CELL ACCUMULATED A WIDE VARIETY OF MUTATIONS THAT HAVE CREATED VARIATION IN THE POPULATIONS.

VIVE LA DIFFÉRENCE!

SOME OF THE MUTATIONS ALLOWED SOME OF THE BACTERIA TO BREED FASTER THAN OTHERS. THE FAST BREEDERS GENERATED MORE OFFSPRING IN EACH GENERATION.

FINALLY, SINCE FASTER BREEDING IS A HERITABLE TRAIT, FAST BREEDERS EVENTUALLY RE-PLACED SLOWER BREEDERS, AND THE ENTIRE POPULATION BECAME FASTER BREEDERS.

WE ARE QUITE FASHIONABLE, NO?

AND SO, AFTER TWO DECADES, THE E. COLI IN ALL TWELVE POPULATIONS NOW BREED 75% FASTER THAN THE ORIGINAL BACTERIA.

An excellent example of natural selection.

YES, BUT IT CAME WITH A PRICE.

THEY COULD GROW FASTER ON THEIR DAILY BIT OF GLUCOSE, BUT THEY BECAME LESS EFFICIENT AT DIGESTING OTHER SUGARS.

Hey, that's a trade-off!

Hmmm, this is starting to sound sus-piciously like a review session, Bloort.

REALLY, SIRE? I HADN'T NOTICED.

137

But what does all of this have to do with squinches, sea cucumbers, and rewinding the tape of life?

AH, YES, THANK YOU FOR GETTING US BACK ON TASK, O FANTASTICALLY FOCUSED ONE.

ONE OF THE MOST EXCITING THINGS TO HAPPEN IN THIS EXPERIMENT IS THAT **ONE** OF THE TWELVE BACTERIAL POPULATIONS **EVOLVED THE ABILITY TO EAT CITRATE.**

NO WAY!

?

What's **CITRATE?**

Beats me.

I just assumed **EXCITEMENT** was the appropriate response.

CITRATE IS A CARBON COMPOUND THAT EACH POPULATION OF *E. COLI* WAS SWIMMING IN.

Literally swimming?

YES, YOUR HIGHNESS. IT WAS IN THEIR WATER. DR. LENSKI PUT IT THERE AS AN ALTERNATIVE FOOD SOURCE TO SEE IF ANY OF THE *E. COLI* WOULD EVOLVE A WAY TO EAT IT.

So, *E. coli* that could digest citrate had more available food than those that couldn't.

Mmm. More food. That's **MY** kind of advantage.

But wait a minute. Did you say the ability to digest citrate evolved in only **ONE** of the twelve populations?

THAT'S CORRECT, YOUR HIGHNESS.

But it's a huge advantage. Why didn't the other populations evolve it, too?

BECAUSE EVOLUTION ISN'T FORWARD-LOOKING, YOUR HIGHNESS. NATURAL SELECTION CAN WORK ONLY ON THE MUTATIONS ON HAND. SOME MAY IMPROVE AN ORGANISM'S ABILITY TO SURVIVE, BUT MOST DO NOT.

And those mutations appear **RANDOMLY**, Dad.

So how did Dr. Lenski **"REWIND"** all of this?

HE AND HIS COLLEAGUES USED TWELVE FOSSIL ANCESTORS FROM THE LINE OF CITRATE-EATING BACTERIA TO GROW NEW POPULATIONS AND SEE IF THEY WOULD EVOLVE THE ABILITY TO EAT CITRATE AGAIN.

THEY ALLOWED THESE NEW POPULATIONS TO BREED FOR THOUSANDS OF GENERATIONS AND THEN LOOKED THROUGH ALMOST 40 TRILLION CELLS FOR *E. COLI* THAT COULD EAT CITRATE.

BUT ONLY A FEW COULD.

So, despite the fact that the ability to eat citrate would be a HUGE advantage for the *E. coli*...the evolution of the citrate-eating trait wasn't inevitable.

THAT IS CORRECT, YOUR HIGHNESS. IT APPEARS CITRATE-EATING REQUIRES AT LEAST **TWO DIFFERENT MUTATIONS** THAT NEED TO HAPPEN IN A PARTICULAR SEQUENCE. SO, NO -- EVEN THOUGH THIS TRAIT WAS AN ADVANTAGE, ITS APPEARANCE WAS NOT INEVITABLE.

It was a fluke!

IN A SENSE, ALL LIFE FORMS ARE FLUKES, MY MOST ASTOUNDINGLY ASTUTE MONARCH. NO TRAIT IS INEVITABLE.

SQUINCHES ARE THE FORTUNATE RESULT OF A SPECIFIC SEQUENCE OF EVENTS ON OUR PLANET, INCLUDING **UNIQUE, RANDOM MUTATIONS** THAT THRIVED UNDER THE UNIQUE ENVIRONMENTAL CONDITIONS IN WHICH OUR ANCESTORS LIVED.

So, are you suggesting, Bloort, that if we were to rewind **THE EVOLUTIONARY HISTORY OF SQUINCHES** and replay it from ancient times, we might have evolved differently?

W-WELL, YES, WE WOULD HAVE BEEN S-SUBJECT TO DIFFERENT MUTATIONS AND SLIGHTLY D-DIFFERENT EVOLUTIONARY PRESSURES, O SUPREMELY SAGACIOUS ONE.

We might have been more like these **BRAINLESS, MUD-EATING SEA CUCUMBERS**?

GULP.

THAT...UM...IS A D-DEFINITE POSSIBILITY, MOST MERCIFUL AND SINGULARLY SPECIAL SOVEREIGN.

Well, Bloort, do you **KNOW** what I have to say to **THAT**?

WH-WHAT?

Whew! **LUCKY US.** I hate eating mud.

AWESOME! And since we evolved brains, we get to experience the wonders of the universe. Like life forms from other planets!

Yes, yes. But we also have **IMPORTANT AFFAIRS** of state to attend to.

Bloort, our people are still facing a **GENETIC CRISIS**, and your research on earth evolution appears to have opened some **EXCITING POSSIBILITIES**.

At first I thought only knowledge of these **HIGHLY EVOLVED HUMANS** would be of use to us. However, your data on Earth evolution shows quite clearly that all life there is intimately related.

Although we can't use any of their alien genes **DIRECTLY**, we might get some **IDEAS** for a cure by looking at the genomes of almost anything from bacteria to hominids.

IF I MAY HUMBLY SUGGEST, YOUR HIGHNESS, YOU COULD ASSEMBLE A TEAM OF SCIENTISTS HERE ON GLARGAL TO BEGIN A STUDY OF OUR OWN EVOLUTIONARY HISTORY WHILE I CONTINUE MY WORK ON EARTH.

What's your plan, Bloort?

I COULD BEGIN A MAJOR **GENETIC STUDY** OF ALL LIFE ON EARTH. THERE COULD BE ANY NUMBER OF **UNIQUE ADAPTATIONS** THAT WE COULD MINE FOR THE GOOD AND GLORY OF YOUR MOST ROYAL PERSONAGE AS WELL AS ALL OF SQUINCHDOM.

ONE THING IN PARTICULAR I WOULD LIKE TO EXAMINE. THE CONSTANT EVOLUTIONARY STRUGGLE BETWEEN ORGANISMS AND **INFECTIOUS VIRUSES AND BACTERIA** MAY BE A FERTILE GROUND FOR ADAPTATIONS THAT MAY BE USEFULLY APPLIED TO OUR OWN **GENETIC MALADY.**

I like it. I like it a lot.

I'll get my people on the job **HERE.** You get your tentacles back to **EARTH.**

Let's go, Son.

Uh...

"Uh?" There is no "uh." I just issued a command. Honestly, Son, you have a lot to learn about being a ruler.

Yeah, Dad, about that... I kinda want to go with Bloort.

Son, Bloort is **VERY** busy.

TRUE, YOUR HIGHNESS, BUT, GIVEN WHAT THE PRINCE HAS LEARNED AND HIS APTITUDE FOR SCIENCE, I THINK HIS ASSISTANCE COULD MAKE THE WORK GO MUCH QUICKER.

It's too danger-ous...

But, Dad, "Danger" is my middle name.

HA-HA, THAT'S VERY --

No, actually it **IS** his middle name, Bloort. I knew it was bad idea at the time, but I couldn't help myself. It just sounded so cool.

Very well. You may go.

PAT!

141

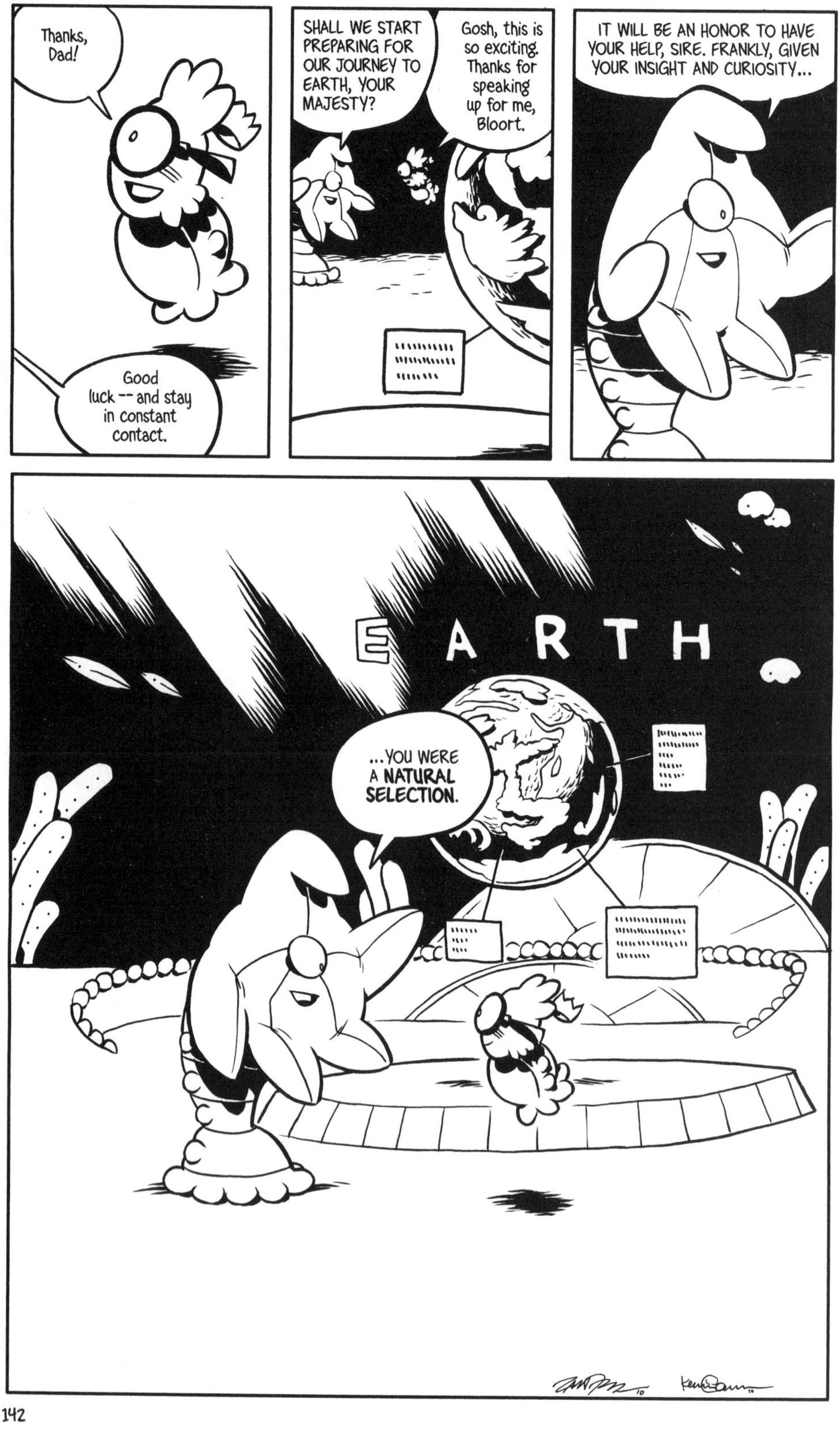
Thanks, Dad!
Good luck -- and stay in constant contact.
SHALL WE START PREPARING FOR OUR JOURNEY TO EARTH, YOUR MAJESTY?
Gosh, this is so exciting. Thanks for speaking up for me, Bloort.
IT WILL BE AN HONOR TO HAVE YOUR HELP, SIRE. FRANKLY, GIVEN YOUR INSIGHT AND CURIOSITY...
EARTH
...YOU WERE A NATURAL SELECTION.

Suggested Reading
Glossary

SUGGESTED READING

PERIODICALS

CURRENT BIOLOGY (www.cell.com/current-biology/home)

DISCOVER (www.discovermagazine.com)

NATIONAL GEOGRAPHIC (www.nationalgeographic.com)

NATURAL HISTORY (www.naturalhistorymag.com)

SCIENCE NEWS (www.sciencenews.org)

SCIENTIFIC AMERICAN (www.SciAm.com)

BOOKS

Carroll, Sean B. *THE MAKING OF THE FITTEST.* New York: W. W. Norton & Company, 2007.

Cochran, Gregory, and Henry Harpending. *THE 10,000 YEAR EXPLOSION.* New York: Basic Books, 2009.

Coyne, Jerry. *WHY EVOLUTION IS TRUE.* New York: Penguin, 2010.

Darwin, Charles. *THE ORIGIN OF SPECIES: THE ILLUSTRATED EDITION.* New York: Sterling, 2008.

Dawkins, Richard. *THE GREATEST SHOW ON EARTH.* New York: Free Press, 2009.

Erwin, Douglas H. *EXTINCTION: HOW LIFE ON EARTH NEARLY ENDED 250 MILLION YEARS AGO.* Princeton, NJ: Princeton University Press, 2006.

Fairbanks, Daniel J. *RELICS OF EDEN.* Amherst, NY: Prometheus Books, 2007.

Fortey, Richard. *TRILOBITE: EYEWITNESS TO EVOLUTION.* New York: Vintage Books, 2001.

Gould, Stephen Jay. *WONDERFUL LIFE.* New York: W. W. Norton & Company, 1990.

Mayor, Adrienne. *THE FIRST FOSSIL HUNTERS.* Princeton, NJ: Princeton University Press, 2000.

Mindell, David P. *THE EVOLVING WORLD.* Cambridge, MA: Harvard University Press, 2006.

Shubin, Neil. *YOUR INNER FISH.* New York: Pantheon, 2008.

Weiner, Jonathan. *THE BEAK OF THE FINCH.* New York: Vintage Books, 1994.

WEBSITES

UNIVERSITY OF CALIFORNIA, BERKELEY
evolution.berkeley.edu

NATIONAL SCIENCE FOUNDATION
www.nsf.gov/news/special_reports/darwin

THE TREE OF LIFE WEB PROJECT
tolweb.org/tree

THE LOOM (a blog by the science writer Carl Zimmer)
blogs.discovermagazine.com/loom

GLOSSARY

ADAPTATION: Any anatomical, physiological, or behavioral trait that gives an organism an advantage in the struggle to survive and pass its genes onto the next generation.

ADAPTATION

ALPHAPROTEOBACTERIA: Primitive energy-producing bacteria that may have been the progenitor of energy-producing mitochondria found in eukaryote cells.

AMINO ACIDS: Organic molecules that are strung together by ribosomes to build proteins. There are twenty amino acids used by the human body.

AMMONOIDS: Extinct relatives of squid, octopuses, and cuttlefish. Most had coiled shells but some had straight, tapered shells that looked like large ice cream cones.

AMMONOID

AMPHIBIANS: Vertebrates that can live on land and in the water but must reproduce in water. Their young breathe though gills like fish, but the adults breathe air. All amphibians must live in moist environments to prevent drying out.

AMYLASE: A protein enzyme that breaks down starch into glucose, the primary sugar used as fuel by most living things.

ANGIOSPERMS: Flowering plants, which first appeared in the Cretaceous Period.

ANTIBIOTIC: A drug that targets and kills bacteria.

ARCHAEA

ARCHAEA: Single-celled prokaryote organisms that have a mix of bacterial and eukaryotic features.

ARDIPITHECUS: Ancient hominid species that lived approximately 4.6 million years ago. It walked on two legs, but had many features that suggest it also spent considerable time in trees.

ARTHROPODS: Invertebrate animals with rigid exoskeletons. They have jointed legs and bodies made up of multiple segments. Modern arthropods include organisms such as crabs, centipedes, spiders, and insects.

ARTIFICIAL SELECTION: The process by which humans selectively breed plants and animals for desired traits.

BACKGROUND EXTINCTIONS: The continuous, low-level loss of species through the normal course of evolution.

BACTERIA: All single-celled organisms that have a cell wall and a single, circular ring of DNA.

BACTERIA

BDELLOIDS: Small multicellular organisms that belong to a large group called rotifers. All bdelloids are female and they reproduce asexually.

CAMBRIAN EXPLOSION: The apparent rapid diversification of multicellular organisms seen in the fossil record from 550 million years ago.

CELL: The basic structural unit of all living things. All cells are composed of a thin membrane filled with a watery substance called cytoplasm.

CENTRAL DOGMA: Theory that states that genetic information in DNA is copied into RNA and that RNA is subsequently decoded to make a protein.

CHORDATES: Animals that have a thin, flexible rod called a notochord running down their backs and a nerve cord right above their notochords.

CHROMOSOME: A threadlike structure of DNA in the cell nucleus that contains multiple genes.

COELACANTH: An evolutionarily ancient species of fish with fleshy, lobe fins. Once thought extinct, living populations of coelacanth have been identified off the coast of east Africa.

CONSTRAINT: An adaptation that limits a species' functional and evolutionary possibilities.

COELACANTH

CONVERGENT EVOLUTION: When two unrelated species independently evolve similar solutions to succeed in a similar environment.

CRUST: The thin, rocky outer layer of the surface of Earth.

CYANOBACTERIA: Bacteria with the ability to photosynthesize sugars by combining sunlight, water, and carbon dioxide. Cyanobacteria produce oxygen as a waste product. They are also referred to as "blue-green algae."

CYTOPLASM: The entire contents of a cell, including the nucleus and organelles suspended in salty water. The cytoplasm is enclosed by the cell's membrane.

DNA: Deoxyribonucleic acid. A double-helical strand of the nucleic acids cytosine, adenine, guanine, and thymine. The unique sequence of these nucleic acids encodes the instructions for making proteins.

ECHINODERMS: The group of organisms containing sea stars and sea cucumbers.

ECHINODERM

ECOSYSTEM: The community that naturally results from the interplay between organisms and the physical aspects of their environment.

ENDOSYMBIOSIS: A mutually beneficial relationship between two organisms in which one is living inside the other. Mitochondria are believed to be the product of an endosymbiosis between eukaryote cells and alphaproteobacteria.

ENDOTHERMY: Meaning "warm inside," endothermy is the ability of an animal to metabolically generate its own internal heat.

ENZYMES: Proteins that run chemical reactions.

EUKARYOTES: Organisms whose cells have several internal membrane-bound chambers, called organelles. One of these chambers, called the nucleus, stores their DNA.

EUKARYOTE

EVOLUTION: A change in the genetic makeup of a population over time.

EXTINCTION: The complete death of a species, with no remaining individuals.

FOSSILS: Any preserved evidence of ancient life. This can include the impression of an organism in

hardened sediment; an organism trapped in amber; or the mineralized hard parts of animals, such as shells and bones.

FUNCTIONAL SHIFT: When a trait adapted for one function starts to be used for a different function.

FUNGI: Multicellular eukaryotes that play an important role as decomposers in an ecosystem. Fungi grow on other organisms— usually dead ones, but not always.

FUNGI

GENE: An individual unit of inheritance composed of DNA and carrying instructions for building proteins.

GENE FLOW: The process by which genes in one gene pool move to another through reproduction.

GENE POOL: The sum of a population's genetic information.

GENOTYPE: The unique set of genes of any individual. It contains the genetic blueprints used to build that individual.

GREAT OXYGENATION: The dramatic increase of atmospheric oxygen 2.2 billion years ago as a result of the activity of cyanobacteria.

GRIFFIN

GRIFFIN:
A mythological animal with the head of an eagle and the body of a lion.

GYMNOSPERMS: Plants with seeds enclosed in a cone, like modern-day pine trees.

HALF-LIFE: The amount of time it takes for half of the unstable atoms in an object to decay.

HEMOGLOBIN: Blood proteins that bind oxygen and transport it in the blood.

HEMOGLOBIN

HOMINID: A member of the group that includes orangutans, chimpanzees, gorillas, and humans.

HOMO HABILIS: Meaning the "handy man," *Homo habilis* is a hominid species that lived 2.3 to 1.4 million years ago.

HOMO HABILIS

HOMOLOGOUS TRAIT: A trait that two or more species inherit from a common ancestor. Such traits can be anatomical, genetic, or developmental.

HOMO SAPIENS: The scientific name for modern humans.

HYBRID: The offspring of two different but related species. Hybrids are usually either nonviable or sterile.

INSECTS: Arthropods with three body segments and six legs. Ninety-eight percent of insect species also have wings.

LACTOSE: The sugar found only in milk.

LATERAL TRANSFERS: The exchange of genetic material between unrelated cells.

MAMMALS: Group of vertebrates with glands producing nutrient-rich fluid called milk, upon which their young feed. Most mammals also have adaptations like hair to keep them warm and sweat glands to cool them off.

MAMMALS

MANTLE: The layer of rock immediately below Earth's crust.

MASS EXTINCTIONS: The simultaneous deaths of an enormous number of species over a geologically short time span.

MUTATION: A change that occurs when a gene's code is copied incorrectly. These copying errors can change the code in one of three ways. Neutral mutations are changes that have no effect on the function of the organism; lethal mutations prevent the successful development of the organism; and beneficial mutations improve the way the organism works.

NATURAL SELECTION: The process by which favorable traits are preserved in a group of organisms and harmful traits die out.

NEOPTERA: Meaning "new wing," they are a group of insects that includes most flying insects. The neoptera have a hinge that makes it possible to fold their wings back when they aren't in use.

PALEOPTERA

NUCLEOTIDES: The building blocks of which DNA and RNA are made. There are five nucleotides: cytosine, adenine, guanine, thymine, and uracil.

PALEOPTERA: Meaning "old wing," they are a group of insects, including dragonflies, that cannot fold their wings back when they are resting.

PANGEA: A gigantic land mass formed when all of Earth's continents came together at the end of the Permian Period, 200 million years ago.

PARACERCEIS SCULPTA: An ocean isopod that lives communally in sponges. Males come in three forms: alpha, beta, and gamma. The alpha is the largest, the beta is smaller in size and looks like a female *Paracerceis sculpta*, and the gamma is the smallest.

PARACERCEIS SCULPTA

PHENOTYPE: The unique physical structure built by an organism's unique set of genes.

PLEIOTROPY: When the change in one trait alters the function of multiple, seemingly unrelated traits.

PRIMATES: The group of mammals that include monkeys, gorillas, chimpanzees, and humans.

PRIMATES

PROKARYOTES: Singled-celled organisms that have a cell wall surrounding their cell membranes; their DNA is a circular strand that floats in the cytoplasm. The cells contain no internal compartments, such as a nucleus, called organelles.

PROTEIN: A molecule composed of a string of smaller molecules called amino acids, connected to each other in a unique sequence.

PROTISTS: Single-celled eukaryotes.

RADIATION: In evolution, the rapid increase in the number of species as a result of new adaptations.

RADIOMETRIC DATING: A method for calculating the age of an object, such as a piece of rock, which uses the half-life of unstable radioactive isotopes.

RIBOSOME: An organelle in the cell that decodes the RNA instructions for a protein and assembles the appropriate amino acids in the appropriate order.

RIBOSOME

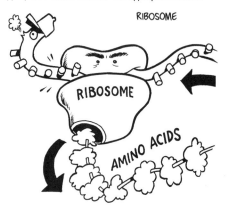

RIBOSOME

AMINO ACIDS

RIBOZYMES: RNA enzymes that can make copies of themselves, run chemical reactions, and store biological information.

RNA: A single strand of nucleic acids that is a copy of a gene and is composed of cytosine, adenine, guanine, and uracil. RNA is sent out of the nucleus and into the cytoplasm, where it is read by a ribosome to make a protein.

SEXUAL SELECTION: The process whereby one sex, usually female, selects for physical or behavioral traits in another sex, usually male. This selection leads to the evolution of traits that gives an individual an advantage in reproducing.

SEXUAL SELECTION

SEXUAL SELECTION

SPECIATION: The process through which new, distinct species evolve from existing species.

STRATA: The rocky layers of Earth's crust. A single layer is called a "stratum."

SYMBIOSIS: When two organisms each benefit by living together with the other.

TETRAPODOMORPH: The group of animals that includes four-legged vertebrate species. In general, this includes all vertebrates except fish.

TIKTAALIK: A species with four leglike limbs that may be a transitional form between water-dwelling fish and land-living vertebrates.

TIKTAALIK

VERNANIMALCULA: The oldest known bilaterally symmetrical organism. Fossils of this animal have been found in Precambrian rock from China.

VERTEBRATES: Animals -- such as fish, amphibians, reptiles, birds, and mammals -- that have a backbone.

VESTIGIAL STRUCTURES: Features retained by organisms, such as leg nubs in some snakes, which appear to have little or no function.

ZYGOTE: The initial single cell that results from the fusion of an egg and sperm. The zygote contains the complete set of genes from both male and female required to grow into an adult organism.

VESTIGIAL STRUCTURE

ABOUT THE SCRIPTER

JAY HOSLER is an associate professor of biology at Juniata College, where he teaches a number of courses, including Evolution, Neurobiology, and Comics and Culture. He has been writing and drawing science comics for more than a decade, and his graphic novels include a biography of a honeybee (*Clan Apis*, 1998) and an action-packed conversation about evolution between Charles Darwin and a follicle living in his left eyebrow (*The Sandwalk Adventures*, 2003). An advocate of comics as a great way to convey the wonder of science, he received funding from the National Science Foundation to develop a comic book textbook on eye evolution called *Optical Allusions* (2008). His comics have won a Xeric Award (1998) and have been nominated for multiple Eisner and Ignatz Awards. Currently, he is the assistant director of Juniata's Center for the Scholarship of Teaching and Learning (SOTL), and his research focuses on the role of comics in improving science education. He lives in central Pennsylvania with his wife and their two little nerdlings.

ABOUT THE ILLUSTRATORS

Cartoonists KEVIN CANNON and ZANDER CANNON (no relation) have been together as a studio since 2004, but their work in comics stretches back to 1993, with such titles as *The Replacement God* (Slave Labor Graphics and Image Comics, 1995), *Top Ten* (America's Best Comics, 1999), and *Smax* (America's Best Comics, 2003). They have also done various work for DC Comics, Dark Horse Comics, the National Oceanic and Atmospheric Administration, and Top Shelf Productions. Their studio work includes the nonfiction graphic novels *Bone Sharps, Cowboys, and Thunder Lizards* (G.T. Labs, 2005), *The Stuff of Life* (Hill and Wang, 2009), and *T-Minus: The Race to the Moon* (Aladdin, 2009). Both Cannons have been nominated for Eisner and Harvey Awards, and Zander has even won a few. They live and work in Minneapolis, Minnesota, and claim to be card-carrying members of the International Cartoonist Conspiracy.